LABORATORY PROCEDURES

FOR THE EXAMINATION

OF

SEAWATER AND SHELLFISH

Fifth Edition 1984

Arnold E. Greenberg
Daniel A. Hunt
Editors

American Public Health Association
1015 Fifteenth Street NW, Washington, D.C. 20005

© 1984 by the American Public Health Association Washington, D.C.

All rights reserved. No part of this publication may be reproduced, stored in a retrieval system, or transmitted in any other form or by any means, electronic mechanical or otherwise without the prior written permission of the publisher.

Library of Congress Cataloging in Publication Data

American Public Health Association.
   Laboratory procedures for the examination of seawater and shellfish.

   Rev. ed. of: Recommended procedures for the examination of sea water and shellfish. 4th ed. 1970.
   Includes bibliographies.
   1. Marine microbiology—Technique. 2. Shellfish—Microbiology—Technique. 3. Poisonous shellfish. I. Greenberg, Arnold E. II. Hunt, Daniel A. III. American Public Health Association. Recommended procedures for the examination of sea water and shellfish.
QR106.A54   1984   594′.028   84-24321

ISBN 0-87553-119-9

84/1M

Committee on Laboratory Standards and Practices, Subcommittee on Laboratory Standards for the Examination of Seawater and Shellfish

ARNOLD E. GREENBERG, *Editor*
Laboratory Services Division, East Bay Municipal Utility District, Oakland, California

DANIEL A. HUNT, *Editor* (retired 1982)

JOHN E. DELANEY, Department of Environmental Quality Engineering, Lawrence Experiment Station, Commonwealth of Massachusetts, Lawrence, Massachusetts

DONALD W. LEAR, U.S. Environmental Protection Agency (retired), Annapolis, Maryland

N. NEUFELD, Inspection Division, Department of Fisheries and Oceans, Burnaby, British Columbia, Canada

NELL C. ROBERTS, Division of Laboratory Services, Office of Health Services and Environmental Quality, State of Louisiana, Lake Charles, Louisiana

MARK D. SOBSEY, Department of Environmental Sciences and Engineering, University of North Carolina, Chapel Hill, North Carolina

TABLE OF CONTENTS

Preface ix

Chapter 1 Apparatus and Media for Bacteriological Examination of Seawater and Shellfish

    1.1 Laboratory Apparatus 1
    1.2 Washing and Sterilization 8
    1.3 Water 8
    1.4 Preparation of Culture Media 9
          A. Storage of Culture Media...................... 9
          B. Adjustment of Reaction 9
          C. Sterilization 10
    1.5 Media and Reagents 10
          A. Media.. 10
          B. Solutions and Reagents 16
    1.6 Bibliography 17

Chapter 2 Field Methodology

    2.1 Introduction 18
          A. Sampling Stations 18
          B. Sampling Frequency 18
          C. Accountability................................ 19
    2.2 Collection of Bacteriological Samples 19
          A. Water Samples 19
          B. Shellfish Samples 20
          C. Bottom Sediment Samples 22
          D. Sample Transportation 23
    2.3 Physical and Chemical Observations 23
          A. Temperature Measurement.................... 24
          B. Salinity 25
          C. Dissolved Oxygen 30
          D. Hydrogen Ion Concentration (pH) 32
          E. Turbidity.................................... 33
          F. Runoff (Stream Flow, Precipitation) 34
          G. Observations 34
          H. List of Suppliers.............................. 34
    2.4 References.. 36

Table of Contents

Chapter 3 Procedures for the Bacteriological Examination of Seawater and Shellfish

- 3.1 Introduction 37
  - A. Multiple Tube Fermentation Technique 37
  - B. Membrane Filter (MF) Technique 38
  - C. Standard Plate Count (Heterotrophic Plate Count) 39
  - D. Coliform Bacteria 39
  - E. Fecal Coliform Group 40
- 3.2 Multiple Tube Fermentation Test for Coliform and Fecal Coliform Bacteria 41
  - A. Sample Collection and Handling 41
  - B. Lauryl Sulfate Tryptose (LST) Fermentation Test 41
  - C. A-1 M Fermentation Test for Fecal Coliform Bacteria 45
  - D. Examination of Shellfish 48
- 3.3 Differentiation of Coliform Organisms 53
  - A. Coliform Verification 53
  - B. Cytochrome Oxidase Test 53
  - C. IMViC Tests 54
- 3.4 Membrane Filter (MF) Method for Fecal Coliform (and Coliform) Bacteria 55
  - A. Sample Collection and Handling 55
  - B. Inoculation Procedure 56
  - C. Incubation 56
  - D. Counting 57
  - E. Computing and Reporting Results 57
- 3.5 Standard Plate Count (Heterotrophic Plate Count) .. 58
  - A. Sample Collection and Handling 58
  - B. Inoculation Procedure 58
  - C. Incubation 59
  - D. Counting Plates and Computing Results 59
- 3.6 References 61
- 3.7 Bibliography 62

Chapter 4 Bioassay Procedures for Shellfish Toxins

- 4.1 Introduction 64
  - A. Paralytic Shellfish Poison 64
  - B. *Ptychodiscus brevis* Toxin(s) 65

4.2	Method for PSP	66
	A. Apparatus	66
	B. Reagents	66
	C. Special Materials	66
	D. Standardization of Bioassay	67
	E. Preparation for Analysis of Samples	71
	F. Mouse Bioassay Test	73
4.3	Method for *Ptychodiscus brevis* Toxin(s)	74
	A. Apparatus	74
	B. Reagents	74
	C. Test Animals	75
	D. Preparation for Analysis of Samples	75
	E. Mouse Bioassay Test	76
4.4	References	78
4.5	Bibliography	79

Chapter 5 Procedures for the Virological Examination of Seawater, Shellfish and Sediment

5.1	Introduction	81
5.2	General Methods and Procedures	82
	A. Sterilizing Apparatus, Materials and Reagents	82
	B. Hydrogen Ion Concentration (pH)	82
	C. Sample Collection	83
	D. Sample Storage	83
	E. Decontamination of Processed (Concentrate) Samples	83
5.3	Enteric Virus Concentration from Shellfish	84
	A. Introduction and General Description	84
	B. Equipment and Materials	87
	C. Reagents	89
	D. Procedures	91
5.4	Enteric Virus Concentration from Sea and Estuarine Water	95
	A. Introduction and General Description	95
	B. Equipment, Materials and Reagents	96
	C. Procedure	100
5.5	Enteric Virus Concentration from Sediments	106
	A. Introduction and General Description	106
	B. Equipment, Materials and Reagents	106

TABLE OF CONTENTS

	C. Procedure	109
5.6	Procedures for Isolating and Quantifying Enteric Viruses in Shellfish, Water, and Sediment Concentrates	110
	A. Introduction...................................	110
	B. Host Systems for Virus Isolation and Assay......	110
	C. Virus Assay Methods	111
5.7	References..	114
5.8	Bibliography	117

Appendix... 121

PREFACE TO FIFTH EDITION

When the Committee on Laboratory Standards and Practices of the American Public Health Association undertook the preparation of the *Compendium of Methods for the Microbiological Examination of Foods* (first edition, 1976), it decided to include a chapter on shellfish. The availability of this book and the related *Standard Methods for the Examination of Water and Wastewater* led the Committee to decide that a fifth edition of *Recommended Procedures for the Examination of Sea Water and Shellfish* would be redundant. However, developments in the shellfish industry and the specific needs of laboratories analyzing only shellfish and seawater indicated that a special publication on shellfish and seawater was, in fact, necessary.

The Committee reviewed the situation, reversed its decision, and invited Daniel A. Hunt of the Shellfish Sanitation Branch of the U.S. Food and Drug Administration to serve as editor for this fifth edition, retitled, *Laboratory Procedures for the Examination of Seawater and Shellfish*. He served as editor and did much of the preliminary manuscript preparation until his retirement in early 1982. Working closely with him was his associate, Dr. John P. Lucas, to whom special thanks are due. In mid-1982, Arnold E. Greenberg assumed the role of editor and brought the book to its final finished state.

It is intended that this book will provide the analytical procedures necessary for the evaluation of shellfish and the waters in which they may be raised. At the 7th National Shellfish Sanitation Workshop in 1971, the U.S. Food and Drug Administration proposed, and the Workshop unanimously accepted, *Recommended Procedures for the Examination of Sea Water and Shellfish* as the official analytical reference of the National Shellfish Sanitation Program. Furthermore, this publication contained the official Food and Drug Administration procedures for the bacterial analysis of domestic and imported molluscan shellfish. This updated Fifth Edition, retitled *Laboratory Procedures for the Examination of Seawater and Shellfish,* will continue to serve the Food and Drug Administration, state shellfish control officers and the industry.

The assistance of the chapter authors and committee members is gratefully acknowledged, as is the invaluable review provided by the U.S. Food and Drug Administration.

CHAPTER 1

APPARATUS AND MEDIA FOR BACTERIOLOGICAL EXAMINATION OF SEAWATER AND SHELLFISH

Nell C. Roberts

1.1 LABORATORY APPARATUS

A. Air Incubators

Incubators shall maintain a uniform and constant internal temperature at all times and must not vary more than ±0.5°C in the areas used. This can be accomplished by using a waterjacketed or anhydric-type incubator with thermostatically controlled low-temperature electric heating units properly insulated and located in or adjacent to walls or floors of the chamber. Preferably equip with mechanical means of circulating air.

Incubators with high-temperature heating units are unsatisfactory because such sources of heat, when improperly placed, frequently cause overheating and excessive drying of media, with consequent inhibition of bacterial growth. Incubators so heated may be operated satisfactorily by replacing high-temperature units with suitable wiring arranged to operate at a lower temperature and by installing mechanical air circulation devices. It is desirable, where ordinary room temperatures vary excessively, to keep laboratory incubators in special rooms maintained at a few degrees below the recommended incubator temperature.

Alternatively, use special incubating rooms well insulated and equipped with properly distributed heating units, forced air circulation, and air exchange ports, provided that they conform to required temperature limits. When such rooms are used, record the daily temperature range in areas where plates or tubes are incubated.

Provide incubators with shelves so spaced as to assure temperature uniformity throughout the chamber. Leave a 2.5-cm space between adjacent stacks of culture dishes and between walls and stacks. Do not stack culture dishes over four high. Maintain an accurate thermometer, traceable to the National Bureau of Standards, at representative locations within the incubator and record daily temperature readings. Immerse thermometer bulb in liquid (glycerine, water, or mineral oil). It is desirable, in addition, to maintain a recording thermometer within the incubator on the middle shelf to record temperature variations over a 24-h period. Frequently record temperature variations within the incubator when it is filled to maximum capacity.

B. Water Bath Incubators

The specificity of the fecal coliform test is related directly to the incubation temperature. A temperature tolerance of $\pm 0.2°C$ is required and can be obtained with most types of water baths equipped with a circulation system and a cover to prevent heat loss. Include culture controls of EC gas-positive *Escherichia coli* and EC-negative *Enterobacter aerogenes* with each set of tests.

For incubating membrane filter cultures, provide for a high level of humidity.

C. Thermometers

Use accurate thermometers regularly checked against a thermometer certified by the National Bureau of Standards. Preferably use thermometers graduated at 0.1°C intervals.

D. Hot-Air Sterilizing Ovens

Use hot-air sterilizing ovens of sufficient size to prevent internal crowding; constructed to give uniform and adequate sterilizing temperatures; and equipped with suitable thermometers capable of registering accurately in the range 160 to 180°C. Optionally use a temperature-recording instrument.

APPARATUS AND MEDIA

E. Autoclaves

Use autoclaves of sufficient size to prevent internal crowding; constructed to provide uniform temperatures within the chambers (up to and including the sterilizing temperature of 121°C); and equipped with an accurate thermometer, the bulb of which is located on the exhaust line so as to register the minimum temperature within the sterilizing chambers. A temperature-recording instrument is optional. Connect pressure gauges and properly adjusted safety valves directly to either saturated steam supply lines or to a suitable steam generator. The autoclave shall be capable of reaching the desired temperature within 30 min.

F. Colony Counters

Use Quebec-type colony counter, dark-field model preferred, or one providing equivalent magnification (1.5 diameters) and satisfactory visibility.

G. pH Equipment

Use electrometric pH meters, accurate to at least 0.1 pH units, for determining pH values of media.

H. Balances

Use balances providing a sensitivity of at least 0.1 g at a load of 150 g, with appropriate weights. Use an analytical balance having a sensitivity of 1 mg under a load of 10 g for weighing less than 2 g of materials.

I. Media Preparation Utensils

Use borosilicate glass or other suitable noncorrosive equipment such as stainless steel. Use glassware that is clean and free of foreign residues. Use metalware not containing toxic materials—copper, zinc, antimony, chromium, or detergents—which might contaminate the media.

J. Pipets

Use pipets of any convenient size, provided that they deliver the required volume accurately and quickly. The error of calibration for a given manufacturer's lot shall not exceed 2.5%. Use pipets having graduations distinctly marked and with unbroken tips. Discard pipets with damaged tips. Pipets conforming to the APHA standards given in *Standard Methods for the Examination of Dairy Products* may be used. Do not use pipets larger than 10 mL to deliver 1 mL. Do not use pipets larger than 1 mL to deliver 0.1 mL.

K. Pipet Containers

Use boxes of aluminum or stainless steel, end measurement 5 to 7.5 cm, cylindrical or rectangular, and length about 40 cm. Paper wrappings may be substituted, provided no deleterious or toxic materials adhere to the pipets. Do not use copper or copper alloy cans or boxes as pipet containers.

L. Dilution Bottles or Tubes

Use bottles or tubes of resistant glass, preferably borosilicate glass, closed with glass or rubber stoppers, or screw caps equipped with liners that do not produce toxic or bacteriostatic compounds on sterilization. Do not use cotton plugs as closures. Mark graduation levels indelibly on side of dilution bottle or tube. Plastic bottles of nontoxic material and acceptable size may be substituted for glass provided they can be sterilized properly.

M. Petri Dishes

For the plate count, use glass or plastic petri dishes about 100 × 15 mm. Use dishes the bottoms of which are free from bubbles and scratches and flat so that the medium will be of uniform thickness throughout the plate. Presterilized plastic petri dishes may be substituted for glass dishes for single use only. Sterilize glass petri dishes and store in metal cans (aluminum or stainless steel, but not copper), or wrap in paper, preferably best-quality sulfate pulp (kraft), before sterilizing.

APPARATUS AND MEDIA

For the membrane filter technique, use loose lid glass or plastic dishes, 60 × 15 mm, or tight-lid dishes 50 × 11 mm. If glass petri dishes are used, take precautions to prevent possible loss of medium by evaporation with resultant change in medium concentration and to maintain a humid environment for optimum colony development.

N. Fermentation Tubes and Vials

Use fermentation tubes of any type, if their design permits conforming to medium and volume requirements for concentration of nutritive ingredients as described subsequently (Chapter 3) and provided they are made of resistant non-toxigenic material. Metal or plastic tube closures are preferred provided that neither volatile nor toxic or bacteriostatic compounds are produced on sterilization. Do not use cotton plugs when the fermentation tube is to be used in the fecal coliform test and incubated in an elevated temperature water bath.

O. Inoculating Equipment

Use wire loops made of 22- or 24-gauge nickel alloy* or platinum-iridium for flame sterilization. Single-service resterilizable transfer loops of aluminum or stainless steel may be used. Use loops at least 3 mm in diameter. Sterilize by dry heat or steam. Single-service disposable hardwood applicators also may be used. Make these 0.2 to 0.3 cm in diameter and at least 2.5 cm longer than the fermentation tube; sterilize by dry heat and store in glass or other nontoxic containers.

P. Sample Bottles

Use bottles of borosilicate glass or other materials resistant to the solvent action of water, of any suitable size and shape, provided that the bottles are capable of being properly washed and sterilized; will contain a sufficient volume of sample for all the required tests plus adequate space to allow for effective shaking; and will maintain samples uncontaminated until the examinations are complete. Ground

---

* Chromel, nichrome, or equivalent.

glass-stoppered bottles, preferably wide-mouthed and of resistant glass, are recommended.

Plastic bottles may be used provided they can be sterilized repeatedly at 121°C for 15 min without distortion and also provided they do not produce toxic or bacteriostatic compounds.

Metal or plastic screw-cap closures with liners may be used on sample bottles provided that no toxic or bacteriostatic compounds are produced on sterilization.

Before sterilization, cover tops and necks of sample bottles with metal foil, rubberized cloth, or heavy impermeable paper.

Q. Filtration Units

The filter-holding assembly (constructed of glass, autoclavable plastic, porcelain, or stainless steel) consists of a seamless funnel fastened to a base bearing a porous plate for support of the filter membrane. The funnel unit should be attachable to the base by means of a convenient locking device. The design should insure the membrane filter to be held securely on the porous plate of the receptacle without mechanical damage and allow all fluid to pass through the membrane during filtration. Separately wrap the two parts of the assembly in heavy wrapping paper. Sterilize by autoclaving, and store until use.

For filtration, mount receptacle of filter-holding assembly in a 1-L filtering flask with a side tube or other suitable device such that a pressure differential can be exerted on the filter membrane. Connect flask to an electric vacuum pump, a filter pump operating on water pressure, a hand aspirator, or other means of securing a pressure differential.

Use sterile filtration units at the beginning of each filtration series as a minimum precaution to avoid accidental contamination. A filtration series is considered to be interrupted when an interval of 30 min or longer elapses between sample filtrations. After such interruption, treat any further sample filtration as a new filtration series and sterilize all membrane filter holders in use. Decontaminate this equipment between successive filtrations by using an ultraviolet (UV) sterilization procedure. A 2-min exposure to UV radiation is sufficient. Do not expose membrane-filter culture preparations to random UV radiation leaks that might emanate from the sterilization cabinet. Eye protection is recommended; either safety glasses or prescription-ground glasses afford adequate eye protection against stray radiation from a UV sterilization cabinet that is not light-tight during the exposure interval.

R. Membrane Filters

Use membrane filters with a rated pore diameter such that there is complete bacterial retention. Use only those filter membranes that have been found, through adequate quality control testing and *certification by the manufacturer,* to exhibit: stability in use, freedom from chemical extractables that may inhibit growth and development of bacteria, a satisfactory speed of filtration, no significant influence on medium pH, and no increase in number of confluent colonies or spreaders compared to control membrane filters. Preferably use membranes grid-marked in such a manner that bacterial growth is neither inhibited nor stimulated along the grid lines. Follow manufacturer's directions for handling and sterilizing filters. Suitable packaged filters designed for autoclave sterilization, or if desired, presterilized filters, are available.

S. Absorbent Pads

Absorbent pads for nutrients are disks of filter paper or other material known to be of high quality and free of sulfites or other substances that could inhibit bacterial growth. Use pads approximately 48 mm in diameter and of such thickness to absorb 1.8 to 2.2 mL of medium. Presterilized absorbent pads or pads subsequently sterilized in the laboratory should release less than 1 mg total acidity (calculated as $CaCO_3$) when titrated to the phenolphthalein end point, pH 8.3, using $0.02N$ NaOH. Sterilize pads simultaneously with membrane filters in resealable kraft envelopes, or separately in other suitable containers. Dry pads so they are free of visible moisture before use.

T. Forceps

Use forceps that are round-tipped without corrugations on the inner sides of the tips. Sterilize before use by dipping in 95% ethyl or absolute methyl alcohol and flaming.

U. Microscope and Light Source

Count colonies on membrane filters using a magnification of 10 to 15 diameters and a light source adjusted to give maximum sheen discernment. Optimally use a binocular wide-field dissecting microscope. However, a small fluorescent lamp with magnifier is acceptable. Use cool white fluorescent lamps. Do not use a microscope illuminator with optical system for light concentration from an incandescent light source for coliform colony identification on Endo-type media.

1.2 WASHING AND STERILIZATION

A. Washing

Cleanse all glassware thoroughly with a suitable detergent and hot water, rinse with hot water to remove all traces of residual washing compound, and finally rinse with laboratory pure water. Establish effectiveness of the rinsing procedure by testing as described in the latest edition of *Standard Methods for the Examination of Water and Wastewater*. Before using a new supply of detergent test for inhibitory residues.

B. Sterilization

Sterilize glassware, except when in metal containers, for not less than 60 min at a temperature of 170°C, unless it is known from recording thermometers that oven temperatures are uniform, under which exceptional condition use 160°C. Heat glassware in metal containers to a temperature of 170°C for not less than 2 h.

Sterilize sample bottles (except plastic) as above or in an autoclave at 121°C for 15 min. Plastic bottles may be sterilized with low temperature ethylene oxide gas. Assure that all gas has been removed from the container before using it.

1.3 WATER

To prepare culture media and reagents, use only distilled or demineralized water that has been tested and found free from traces

APPARATUS AND MEDIA

of dissolved metals and bactericidal or inhibitory compounds. Test for bactericidal and inhibitory compounds periodically by the procedure outlined in the latest edition of *Standard Methods for the Examination of Water and Wastewater*.

1.4 PREPARATION OF CULTURE MEDIA

A. Storage of Culture Media

Store culture media in any clean, dry space where excessive evaporation and danger of contamination have been eliminated.

Liquid media in fermentation tubes, if stored at refrigeration or even moderately low temperatures, may dissolve sufficient air to produce, upon incubation at 35°C, a bubble of air in the tube. Incubate fermentation tubes that have been stored at a low temperature overnight before use and discard those tubes containing air.

Fermentation tubes may be stored at approximately 25°C; but because evaporation may proceed rapidly under these conditions—resulting in marked changes in concentration of the ingredients—do not store at this temperature for more than 1 week. Use of screw-capped fermentation tubes or refrigerator storage can extend the storage period up to 3 months.

B. Adjustment of Reaction

Express reaction of culture media as pH. The increase in the hydrogen ion concentration (decrease in pH) during sterilization will vary slightly with the individual sterilizer in use and the initial reaction required to obtain the correct final reaction will have to be determined. The decrease in pH usually will be 0.1 to 0.2 but occasionally may be as great as 0.3. When buffering salts such as phosphates are present in the medium, the decrease in pH value will be negligible.

Make tests to control pH with a pH meter and at 25°C. Calculate volume of approximately $1N$ NaOH solution to add to the bulk medium to reach the prescribed reaction. After adding and mixing thoroughly, check reaction and adjust if necessary. The required final pH is given in the directions for preparing each medium. If a specific pH is not prescribed, adjustment is unnecessary. Keep record of pH.

LABORATORY PROCEDURES FOR SEAWATER AND SHELLFISH

C. Sterilization

Sterilize all media according to the manufacturer's instructions unless otherwise stated. Most specifications require sterilization in an autoclave at 121°C for 15 min after the temperature has reached 121°C. When the pressure reaches zero, remove medium from autoclave and cool quickly to avoid decomposition. To permit uniform heating and rapid cooling, pack materials loosely and in small containers. The maximum elapsed time for exposure of sugar broths to any heat is 45 min.

1.5 MEDIA AND REAGENTS

The need for uniformity dictates the use of dehydrated media.

This section contains formulas for all culture media, solutions, and reagents referred to in Chapter 3. For ease of reference, media and other solutions and reagents are listed alphabetically within each group.

Maintain written quality control records on preparation of media and reagents including results of productivity and inhibition tests, pH, sterilization time and temperature, and other pertinent data.

A. Media

1. A-1 broth

Lactose	5.0 g
Tryptone	20.0 g
Sodium chloride, NaCl	5.0 g
Salicin	0.5 g
Polyethylene glycol *p* isooctylphenyl ether*	1.0 mL
Distilled water	1 L

Heat to dissolve solid ingredients, add ether, and adjust to pH 6.9 ± 0.1. Dispense 10 mL into 150 × 18 mm tubes for inoculum of 1 mL or less. For 10 mL inocula, prepare double strength medium and dispense 10 mL into 175 × 22 mm tubes. Sterilize by autoclaving

---

* Triton X-100, Rohm and Haas Co., or equivalent.

for 10 min at 121°C. A floculant precipitate is common but it does not impair performance.

This medium must be made from the individual ingredients specified; acceptable preformulated, dehydrated medium currently is unavailable.

2. *Brilliant green lactose bile broth, 2%*

Peptone or gelysate	10.0 g
Lactose	10.0 g
Oxgall	20.0 g
Brilliant green	0.0133 g
Distilled water	1 L

Distribute in fermentation tubes and sterilize at 121°C for 15 min. pH after sterilization should be between 7.1 and 7.4.

3. *Buffered glucose broth (MR-VP)*

Polypeptone or equivalent	5.0 g
Glucose	5.0 g
Dipotassium hydrogen phosphate, K_2HPO_4	5.0 g
Distilled water	1 L

Distribute 10-mL portions in culture tubes and sterilize by autoclaving at 121°C for 12 to 15 min, providing that the total time of exposure to heat is not longer than 30 min.

4. *EC medium*

Tryptose or trypticase	20.0 g
Lactose	5.0 g
Bile salts mixture, or bile salts No. 3	1.5 g
Dipotassium hydrogen phosphate, K_2HPO_4	4.0 g
Potassium dihydrogen phosphate, KH_2PO_4	1.5 g
Sodium chloride, NaCl	5.0 g
Distilled water	1 L

Final pH should be 6.9 after sterilization.

5. *M-FC broth*

Broth Base

Tryptose or biosate	1.0 g

Proteose peptone No. 3 or polypeptone .... 0.5 g
Yeast extract .... 0.3 g
Sodium chloride, NaCl......................... 0.5 g
Lactose .... 1.25 g
Bile salts .... 1.5 g
Aniline blue (water blue) .... 0.01 g
Distilled water .... 100 mL
Rosolic Acid Solution
Rosolic acid .... 1.0 g
Sodium hydroxide, NaOH, 0.2N .... 100 mL

(Rosolic acid salt solution may be stored for 2 weeks in the dark in a refrigerator. Discard if it changes from a red to muddy brown color.)

Suspend the M-FC broth base in 100 mL distilled water. Add 1 mL rosolic acid salt solution. Heat to boiling (*do not autoclave*). Cool and use. Final pH should be 7.4 ± 0.1. (Note: 1.5 g agar may be added before boiling to solidify this medium.)

6. Indole test medium (Tryptophane broth)

Tryptone or trypticase, or equivalent .... 10.0 g
Distilled water .... 1 L

Add tryptone to cold distilled water. While stirring, heat to dissolve. Distribute in 5-mL portions in culture tubes and sterilize by autoclaving at 121°C for 15 min.

7. Lactose broth

Beef extract .... 3.0 g
Peptone or gelysate............................ 5.0 g
Lactose .... 5.0 g
Distilled water .... 1 L

pH should be between 6.8 and 7, preferably 6.9. When fermentation tubes for the examination of 10-mL or 100-mL sample portions are prepared, make lactose broth of such strength that adding that volume of sample to the medium will not reduce ingredient concentrations below those of the standard medium. The proper concentration of ingredients is obtained by using the amounts of dehydrated product shown in the following tabulation:

Apparatus and Media

Inoculum mL	Medium in Tube mL	Medium and Inoculum mL	Dehydrated Lactose Broth Required g/L
1	10 or more	11 or more	13.0
10	10	20	26.0
10	20	30	19.5
100	50	150	39.0
100	35	135	50.1
100	20	120	78.0

8. Lauryl tryptose broth (Lauryl sulfate tryptose broth)

Tryptose or biosate	20.0 g
Lactose	5.0 g
Dipotassium hydrogen phosphate, K_2HPO_4	2.75 g
Potassium dihydrogen phosphate, KH_2PO_4	2.75 g
Sodium chloride, NaCl	5.0 g
Sodium lauryl sulfate	0.1 g
Distilled water	1 L

Final pH should be approximately 6.8. Use single strength medium for inocula of 1 mL or less. In examining 10-mL or 100-mL sample portions, maintain proper concentration of the lauryl sulfate medium by using the amounts of dehydrated product shown in the following tabulation:

Inoculum mL	Medium in tube mL	Medium and Inoculum mL	Dehydrated Medium Required g/L
1 or less	10	11	35.6
10	20	30	53.4
10	30	40	47.3
100	35	135	137.1
100	50	150	106.8

9. LES Endo agar

Dissolve the following in 1 L cold distilled water to which 20 mL 95% ethyl alcohol have been added:

Yeast extract ... 1.2 g
Casitone or trypticase ... 3.7 g
Thiopeptone or thiotone ... 3.7 g
Tryptose ... 7.5 g
Lactose ... 9.4 g
Dipotassium hydrogen phosphate, K_2HPO_4 ... 3.3 g
Potassium dihydrogen phosphate, KH_2PO_4 ... 1.0 g
Sodium chloride, NaCl ... 3.7 g
Sodium desoxycholate ... 0.1 g
Sodium lauryl sulfate ... 0.05 g
Sodium sulfite ... 1.6 g
Basic fuchsin ... 0.8 g
Agar ... 15.0 g

Heat to boiling to complete solution. Cool to 45 to 50°C and dispense in 4-mL quantities into the lower section of 60-mm glass or plastic petri dishes. If dishes of any other size are used, adjust volume to give an equivalent depth. Store in the dark at 2 to 10°C for not more than 2 weeks. Do not expose to direct sunlight.

10. *Levine's eosin methylene blue agar*

Peptone or gelysate ... 10.0 g
Lactose ... 10.0 g
Dipotassium hydrogen phosphate, K_2HPO_4 ... 2.0 g
Agar ... 15.0 g
Eosin Y ... 0.4 g
Methylene blue ... 0.065 g
Distilled water ... 1 L

Adjustment of pH not necessary. Sterilize at 121°C for 15 min.

11. *Modified MacConkey agar*

	Single Strength
Peptone	17.0 g
Polypeptone	3.0 g
Lactose	10.0 g
Bile salts No. 3	0.75 g
Agar	13.5 g
Neutral red	0.03 g

Crystal violet .... 0.01 g
Distilled water 1   L

Dissolve ingredients in distilled water, boil for 10 min, and hold in water bath at 45 to 50°C. Use within 6 h. For double strength agar, double the quantities of solid ingredients added per liter.

12. Nutrient agar

Beef extract .... 3.0 g
Peptone .... 5.0 g
Agar .... 15.0 g
Distilled water 1   L

Final pH should be 6.8 ± 0.1.

13. Plate count agar

Tryptone or trypticase .... 5.0 g
Yeast extract .... 2.5 g
Glucose .... 1.0 g
Agar .... 15.0 g
Distilled water 1   L

Add dehydrated medium to cold distilled water in a suitable container. Let soak 3 to 5 min and then dissolve mixture by boiling carefully while stirring frequently or by exposing to actively flowing steam. Dispense into bottles, tubes, or flasks and sterilize at 121°C for 15 min. Final pH should be 7.0 ± 0.1.

14. Simmon's citrate agar

Magnesium sulfate, $MgSO_4 \cdot 7H_2O$ .... 0.2 g
Ammonium dihydrogen phosphate, $NH_4H_2PO_4$ ... 1.0 g
Dipotassium hydrogen phosphate, K_2HPO_4 .... 1.0 g
Sodium citrate, $Na_3C_6H_5O_7 \cdot 2H_2O$ .... 2.0 g
Sodium chloride, NaCl .... 5.0 g
Agar .... 15.0 g
Bromthymol blue .... 0.08 g
Distilled water 1   L

Suspend ingredients in cold distilled water and heat to boiling to dissolve. Distribute in tubes or flasks and autoclave at 121°C for 15 min. Cool medium in slanting position. Final pH should be 6.8.

LABORATORY PROCEDURES FOR SEAWATER AND SHELLFISH

B. Solutions and Reagents

1. Buffered dilution water

a. Stock phosphate buffer solution

Potassium dihydrogen phosphate, KH_2PO_4 .... 34.0 g
Distilled water
Sodium hydroxide, NaOH, $1N$

Dissolve potassium phosphate in 500 mL distilled water. Adjust to pH 7.2 with $1N$ NaOH and make up to 1 L with distilled water.

b. Magnesium chloride solution

Dissolve 81.1 g $MgCl_2 \cdot 6H_2O$ per liter distilled water.

c. Final phosphate buffered dilution water

Stock phosphate buffer solution .... 1.25 mL
Magnesium chloride solution .... 5.0 mL
Distilled water 1   L

Fill dilution bottles or tubes with dilution water so that after sterilization by autoclaving at 121°C for 15 min they will contain the quantity desired with a tolerance of ±2%.

2. Peptone water

Peptone or gelysate............................. 1.0 g
Distilled water 1   L

Dissolve peptone in distilled water and dispense to provide 99 ± 2.0 mL or 9 ± 0.2 mL after autoclaving at 121°C for 15 min.

3. Indole test reagent

Paradimethylaminobenzaldehyde, m.p. 74–75°C .. 5.0 g
Isoamyl alcohol, $CH_3(CH_2)_4OH$ .... 75 mL
Hydrochloric acid, HCl, conc .... 25 mL

Dissolve paradimethylaminobenzaldehyde in isoamyl alcohol and add conc HCl. The final solution should have a light yellow color. Some brands of paradimethylaminobenzaldehyde are unsatisfactory while others become unsatisfactory on aging. If isoamyl alcohol is unavailable, normal amyl alcohol may be used. The amyl alcohol solution should have a pH <6.0. Purchase alcohol and benzaldehyde in as small amounts as consistent with the volume of work to be done. Check new lots of reagents with known positive and negative cultures.

4. **Methyl red indicator solution**

   Methyl red.................................... 0.1 g
   Ethyl alcohol, 95% .... 300 mL

   Dissolve methyl red in alcohol and make up to 500 mL with distilled water.

5. **Voges-Proskauer reagents**

   a. alpha-Naphthol solution

   Naphthol, purified, m.p. 92.5°C, or higher .... 5.0 g
   Absolute ethyl alcohol, C_2H_5OH .... 100 mL

   Prepare fresh solution for each day of use.

   b. Potassium hydroxide solution

   Potassium hydroxide, KOH .... 40 g
   Distilled water .... 100 mL

1.6 BIBLIOGRAPHY

AMERICAN PUBLIC HEALTH ASSOCIATION. 1985. Standard Methods for the Examination of Dairy Products, 15th ed., APHA, Washington, D.C.

AMERICAN PUBLIC HEALTH ASSOCIATION. 1981. Standard Methods for the Examination of Water and Wastewater, 15th ed., APHA, Washington, D.C.

HOROWITZ, W., ed. 1980. Official Methods of Analysis of the Association of Official Analytical Chemists, 13th ed., AOAC, Washington, D.C.

INHORN, S. L., ed. 1978. Quality Assurance Practices for Health Laboratories. American Public Health Association, Washington, D.C.

SPECK, M. L., ed. 1976. Compendium of Methods for the Microbiological Examination of Foods. American Public Health Association, Washington, D.C.

CHAPTER 2

FIELD METHODOLOGY

Donald W. Lear

2.1 INTRODUCTION

The procedures in this chapter describe the collection of samples for bacteriological analyses and the physical or chemical observations or tests that might be required. Physical and chemical data are necessary in interpreting bacteriological data. While it is possible to provide specific details regarding the collection and handling of samples, specifications on the location of sampling stations or on the frequency of sampling cannot be given. It is important to note, however, that the ultimate assessment of a shellfish-raising area depends on a meaningful sampling program supported by appropriate field observations and measurements.

A. Sampling Stations

Locate sampling stations to reflect the desired characteristics of the environment being evaluated. Consider pollution sources and types and local hydrographic conditions such as vertical or horizontal stratification, eddies, and gyres.

B. Sampling Frequency

Select a sampling frequency to reflect the variability of pollution conditions in the environment. For example, consider seasonal variations because greater sampling frequency may be necessary during harvesting periods or high-flow conditions. Runoff can cause greater variability of water quality in a small watershed than in a large

FIELD METHODOLOGY

embayment; therefore, adjust sampling frequencies, in part, by the size of the watershed being studied.

C. Accountability

Thoroughly identify sample containers and field record sheets at all stages of analysis, from field to laboratory. Record name of sampler and routinely provide sample accountability ("chain of custody") from field to laboratory personnel. The chain of custody may be important for quality control or for epidemiological or legal purposes.

Insure that all field observations include date, time, and place of collection; identity of sampler; weather and water conditions at time of collection; and comments on unusual conditions or events observed.

2.2 COLLECTION OF BACTERIOLOGICAL SAMPLES

All sampling personnel should be aware fully of the concept of "aseptic technique" and the integrity of sterile systems. It is imperative, especially under primitive field conditions, that precautions be taken against bacterial contamination or cross-contamination of containers and samples.

A. Water Samples

Collect water samples for bacteriological examination in clean sterile bottles or containers (see Chapter 1). Protect the container fully against contamination before, during, and after collection. Foil wrapping over bottle stoppers before sterilization aids in preventing contamination.

1. *Surface samples*

Surface samples may be collected without the aid of special sampling devices. Keep the sample container unopened until immediately before filling. During sampling protect the container stopper or closure from contamination. In collecting the sample, hold the bottle near the base and plunge it neck downward below the surface. Then tilt it with the neck pointing slightly upward, and during filling, push the container horizontally forward in a direction away from the hand to avoid contamination. To facilitate shaking the sample in the laboratory, do not fill the container completely full.

2. Subsurface samples

A subsurface sampler should be of simple mechanical design and able to hold a sterile container that is nonmetallic and nonbactericidal, with mechanical devices to open and shut the container closure at the desired sampling depth. Keep the container sealed until the sampling depth is reached.

A number of subsurface sampling devices are available commercially. The capillary tube water sampler or hinge samplers with presterilized plastic bags in general use in oceanographic work may be utilized. Specific designs for several samplers have been described.[1]

B. Shellfish Samples

In collecting and submitting shellfish samples, provide an accompanying history and description of the shellfish including the date, time, and place of collection; the area from which the shellfish were harvested; the date and time of harvesting; and the conditions of storage between harvesting and collection. This information may not be obtainable for market shellfish samples. In such cases, identify the shipper, the date of shipment, and the harvesting area as well as the date, time, and place of collection.

Identify individual containers of shellfish and use the same identifying mark on the descriptive form which accompanies the sample.

1. *Collection of shellstock*

a. Equipment and permits

Methods for collecting shellfish vary with species and locality. Generally, local watermen or sportsmen have developed the most economical, efficient, and legal gear for commercial shellfish. Various types of sampling equipment have been described.[2-4] Samples may be collected with rakes, tongs or dredges, or by other means.

Where appropriate, obtain special permits or licenses from the state agency involved for the sampling of shellfish. Most states have specific regulations that must be observed.

FIELD METHODOLOGY

b. Sample containers

Collect samples of shellfish in clean, sterile containers. Use waterproof containers that are durable enough to withstand the cutting action of the shell during transportation. Tin cans with lids, waterproof paper bags, waxed cardboard cups, or plastic bags are suitable.

Keep shellstock samples in dry storage at 10°C or lower, or on ice, until examined. Do not let shellstock contact the ice directly. Do not freeze.

c. Sample size

The size of the sample is governed by the shellfish species being examined. In general, 12 shellfish are adequate; this allows for the selection of 10 sound animals suitable for shucking. For most species this sample size will yield approximately 200 g of meats and shell liquor. Fewer shellfish per sample may be taken where laboratory data have demonstrated that equivalent results are obtained. When shellfish of a smaller size, such as mussels, are being examined, collect more animals to produce the same weight of shucked meats.

With some species of shellfish, such as the Pacific oyster, *Crassostrea gigas,* or the surf clam, *Spisula solidissima,* 10 animals produce far more than 200 g of shucked meats. In such case use appropriate fractions of several animals.

2. Fresh shucked shellfish

A sterile widemouthed jar of a suitable capacity with a watertight closure is an acceptable container for samples of shucked shellfish taken in shucking houses or repacking establishments or for bulk shipments in the market. Transfer the shellfish to the sample jar with a sterile forceps or spoon. Samples of the final product of shucking houses or repacking establishments may be taken in the final packing cans or containers. Follow the appropriate directions for shellstock sample size. Consumer-sized packages also are acceptable for examination provided they contain about the required weight or volume. One appropriately sized package constitutes one sample. Refrigerate samples of shucked shellfish immediately after collection by packing in crushed ice; keep in ice until examined but do not freeze.

3. Frozen shucked shellfish

Consumer-sized packages are acceptable as samples. Samples from larger blocks may be taken by coring with a suitable instrument or by quartering, using sterile technique. Transfer cores or quartered samples to sterile widemouthed jars for transportation to the laboratory.

Keep samples of frozen shucked shellfish frozen at temperatures close to those at which the commercial stock was maintained. When this is not possible, pack samples of frozen shucked shellfish in crushed ice and hold them until examined.

C. Bottom Sediment Samples

While bottom sediments are not routinely used for the evaluation of shellfish areas, they may contain a more persistent record of contamination. This is due to the capacity of sediment particles to adsorb bacteria, sometimes extending bacterial survival in the environment, as well as to the lesser mobility of sediments with respect to water.

1. Sampling equipment

Bacteria and viruses of interest in sanitary microbiology will be almost entirely at the sediment-water interface. A sampler that retrieves a nearly undisturbed interface sample is necessary. Some samplers cause the sediment to fold and become distributed, resulting in dilution of the interface and biasing the population estimates. Other samplers, such as the Van Donsel-Geldreich sampler,[1] have been designed to collect interface samples aseptically.

Corers and large grab samplers such as the Smith-McIntyre sampler generally give fairly undisturbed samples. As it is usually impractical to sterilize such an apparatus in the field between sampling, take precautions to keep the gear as clean as possible between samples.

2. Sampling procedure

For most work in estuarine water, hand held or gravity corers are the most practical samplers. Plastic core liner tubing of 37 mm diam usually gives little distortion of sediment structure. Sterilize or disinfect core liners before use.

Place a core liner in the core barrel of the sampler; secure the liner with the core cutter or core catcher if necessary. Upon retrieval, keep the corer upright, remove the core liner, and immediately cap both ends with sterilized or disinfected core liner caps. If samples are to be transported in core liners to the laboratory for further manipulation, cap and transport upright on ice or under proper refrigeration.

Extrude the sample from the core liner by a snug-fitting piston pushed upward from the bottom of the core until the interface is at the top of the core liner; then place segments of the core, usually 3 to 5 cm long, in a suitable sterile container for further laboratory manipulation. A laboratory rubber stopper of suitable size to fit just into the core liner, mounted on a rod or dowel, serves well as an extruding piston. Extruding can be done in the field, using aseptic technique, or in the laboratory.

In soft muddy bottoms, corers may penetrate too readily to retrieve a satisfactory sample. For soft bottoms an Ekman grab sampler often is satisfactory. Carefully and aseptically remove sample portions from the sediment-water interface in the sampler.

D. Sample Transportation

Transport samples to the laboratory in an insulated container. Pack them to prevent leakage and breakage; cover with crushed or flaked ice to maintain temperatures of 10°C or less but do not freeze. Frozen packs may be used as a coolant. Prevent samples from being submerged in melted ice and do not permit samples to contact the ice directly.

Examine bacteriological samples of seawater immediately after collection, preferably within 1 h. If analysis is delayed, hold samples at a temperature of 10°C or less but do not freeze. Do not examine any seawater sample held more than 30 h after collection.

Examine all nonfrozen bacteriological samples of shellfish or sediment within 6 h after collection; do not examine any shellfish or sediment sample held more than 24 h after collection. Record in the laboratory report the time elapsed between collection and examination.

2.3 PHYSICAL AND CHEMICAL OBSERVATIONS

Physical and chemical observations of the environment are necessary for complete evaluation of shellfish-raising areas. The characteristics of most common concern include air and water temperatures, salinity (conductivity), dissolved oxygen, pH, turbidity (or light extinction), runoff (river flows and precipitation), and observations of unusual occurrences.

Samples usually are grab samples and they often are analyzed instrumentally, or measurements are made *in situ* with probes, or by a combination thereof. Reliable electronic field instrumentation offers a practical capability for many more field observations than grab samplers do. Precalibrate such instruments in the laboratory for accuracy, sensitivity, drift, etc. Supplement field observations obtained electronically with occasional grab samples, analyzed by conventional techniques, as quality control checks. *In situ* probes are especially useful in defining stratified conditions.

When sampling or making observations from a vessel, sample as far as possible from discharge outlets, and if under way, collect samples from the forward part of the vessel to avoid disturbance from the wake. When lying to, take *surface* samples from the side toward which the vessel is drifting, to minimize vessel contamination. This often is impractical with probes because the wires drift under the vessel; therefore, make such observations while drifting away from the wire, but note the wire angle and correct it if necessary.

A. Temperature Measurement

Air and water temperatures may be measured with conventional glass laboratory thermometers, electronic thermometers, reversing thermometers, or continuously recording thermographs. For any type, calibrate the instrument with a National Bureau of Standards certified thermometer. Make comparisons at several temperatures at and between the high and low extremes expected. Measure water temperature at the surface and near the water-sediment interface.

1. *Glass thermometer*

Use a thermometer with a scale marked for every tenth of a degree and etched on the glass of the capillary. The thermometer should be of small thermal capacity to attain rapid equilibrium. In surface "bucket" samples, measure temperature immediately with the thermometer properly immersed in the sample. Subsurface samples may

Field Methodology

be collected in an insulated bottle and read similarly. Armor cases for glass thermometers are extremely useful in the field.

2. *Electronic (thermistor) thermometers*

Electronic (thermistor) thermometers can have a meter or a digital readout. Obtain sufficient length of wire for normal field observations. Check the instrument for accuracy and for tendency to drift.

3. *Reversing thermometers*

Reversing thermometers normally are used only for very accurate oceanographic work. They are quite expensive and require precise calibration and fairly tedious corrections of the readings.[5]

4. *Thermographs (bathythermographs)*

Thermographs (bathythermographs) record temperature as a function of depth, using a pressure transducer in conjunction with a thermal sensor to scribe a thermal profile on a special slide. The slide is read in a calibrated reader. Generally these are used in deep marine environments rather than estuaries. Expendable models as well as glass slide models are available, but cost may be a consideration for routine use.[5]

B. Salinity

Salinity of seawater is defined as the total amount of solid material, in grams, contained in one kilogram of seawater when all the carbonate has been converted to oxide, the bromine and iodine replaced by chlorine, and all organic matter completely oxidized. Salinity may be important in shellfish management and sanitation. It reflects runoff in estuaries where changes in salinity may be more important than concentrations.

Precise titrimetric or electronic methods are available for oceanographic measurements where subtle differences are important. In estuaries, because of runoff and tidal changes, salinity may fluctuate widely and such precise results will be irrelevant. The titrimetric method is included below for standardization purposes. Salinity can be determined indirectly from the index of refraction, from density, or from conductivity. Conductivity at a measured temperature may be used to monitor environmental changes without calculating salinity.

1. Hydrometer method

Salinity measurement by the density (hydrometer) method is simple and reliable. It is well suited for shoreline and small boat observations. The precision is ±0.1‰ (parts per thousand, g/kg). Because hydrometers are precalibrated, or need to be calibrated only once, this is the method of choice.

   a. Apparatus
      i. Hydrometer jar: Use a special jar approximately 400 mm high with inside diameter 45 mm, a rubber-stoppered transparent plastic tube with the same dimensions, or a 500 mL graduated cylinder.
      ii. Thermometer: Graduated in 0.2°C divisions.
      iii. Seawater hydrometer: Use a set of three with specific gravity ranges of 0.996 to 1.011, 1.010 to 1.021, and 1.020 to 1.031. Hydrometer divisions should be 0.002. Have a set calibrated by the National Bureau of Standards for specific gravity of NaCl solutions at 15/4°C.

   b. Procedure

      Fill hydrometer jar ⅔ full of sample and, while holding jar vertically, insert appropriate hydrometer and thermometer. Read and record temperature. Read and record specific gravity after the hydrometer becomes stable; estimate the fourth decimal place.

   c. Calculations
      i. Make temperature correction for specific gravity reading from factors given in Table 1 (see Appendix).
      ii. Determine salinity from Table 2 (see Appendix). Locate corrected density and read salinity from opposite column. Report salinity in parts per thousand, ‰ (g/kg).

2. Induction salinometers

Induction salinometers use an electromagnetic field induced in a coil and determine sample conductivity. At known temperatures, conductivity is converted readily to salinity by the use of tables. Some instruments have an electronic circuit in conjunction with the thermistor to permit conversion so that the readout is expressed as salinity. This type of instrument can be very reliable, has the advantage of simplicity of observation for making many field measurements, can be subcalibrated in the field with a known resistor, is not sensitive to poisoning of electrodes, and is generally a rugged field instrument. Check periodically against standard seawater or the titrimetric method.

3. Electrode salinometers

Electrode salinometers measure conductivity and are manipulated similarly as are induction salinometers. Take care to prevent poisoning or coating of the electrodes. Calibrate the instrument with standard seawater or by the titrimetric method.

4. Refractometers

Salinity can be determined rapidly in the field or laboratory from the refractive index of a very small sample in a special salinity refractometer. The relatively high cost of these instruments may be a factor against selection of this method.

5. Titration

This is the reference method and can be used routinely or for occasional calibration of other methods. The method depends on the relationships of dissolved ions in the open sea and may not be entirely accurate in some estuaries with river inputs of ions in atypical proportions. It is the best practical and reasonably accurate method. Use it to calibrate other methods.

    a. Sampling

        Collect a sample in a 240 mL glass bottle with a No. 6 cork stopper. Pretreat stopper by soaking in melted paraffin wax for 30 to 40 s, draining, and drying. Remove excess wax. (Glass stoppered bottles tend to "freeze" when used with seawater.) To collect sample, rinse bottle three times with water being sampled and fill bottle to shoulder. Seal bottle by forcing waxed cork below level of neck. If samples are not examined within 2 d, dip neck of bottle in melted wax.

The sealed sample is stable indefinitely. Titrate unsealed samples within a few minutes of collection; do not hold unsealed samples more than 1 h before analysis.

b. Apparatus

Automatic zero-adjusting buret. Lubricate, if necessary, with paraffin stopcock grease, never silicone.

c. Reagents

i. Standard seawater: Standard seawater of known chlorinity ("Eau de Mer Normale") is available from the I.A.P.S.O. Standard Sea Water Service, Brook Road, Wormley, Godalming, Surrey, England. Some oceanographic supply houses stock this standard seawater. A secondary standard may be prepared by filtering seawater (chlorinity about 18 g/kg) collected from the open ocean at a depth of at least 50 m. Stabilize with a few crystals of thymol and seal in sample bottles. Use the mean of 10 or more sample titrations as the chlorosity (20°C) of this secondary standard.

ii. Silver nitrate solution, approximately $0.28N$: Dissolve 48.5 g $AgNO_3$ in 500 mL distilled water and dilute to 1000 mL. Store in glass-stoppered brown glass bottle at room temperature.

iii. Potassium chromate indicator solution: Dissolve 63 g K_2CrO_4 in 100 mL distilled water. Add a few drops of $0.28N$ $AgNO_3$ until a definite red precipitate persists. Let stand to settle, filter, and store in a glass dropping bottle.

iv. Standard sodium chloride: Dry about 35 g NaCl to constant weight. Cool and weigh out 29.674 g. Dissolve in distilled water and dilute to 1000 mL. Check this standard against standard seawater and periodically against the secondary seawater standard.

Standardization: Place 0.25 mL standard NaCl solution in a 150 mL erlenmeyer flask. Add 6 drops of chromate indicator and titrate with $AgNO_3$ solution in yellow light until a red precipitate just forms. Stopper flask with a rubber stopper and shake vigorously to break curds of

AgCl. Wash down stopper and continue titration to brown end point. Be consistent in end-point recognition. Repeat titration at least five times and use the average volume in calculating normality.

$$\text{Normality of AgNO}_3 = \frac{12.69}{\text{mL AgNO}_3}$$

d. Procedure

Let sample and AgNO$_3$ titrant come to the same temperature. Use a 25.0 mL sample and titrate as directed above.

e. Calculation

  i. Calculate the chlorosity equivalent of 1 mL AgNO$_3$ solution:

  $$ClEq = N \times 0.0355$$

  where:
  $ClEq$ = chlorosity equivalent, and
  N = normality of AgNO$_3$.

  ii. Calculate the chlorosity:

  $$Cl_0 = d \times ClEq \times \frac{1000}{25}$$

  where:
  d = mL titrant used, and
  $ClEq$ = chlorosity equivalent AgNO$_3$.

  iii. Convert chlorosity to salinity by using Table 3 (see Appendix). Record salinity, S, as ‰. For example, if 47.23 mL titrant (0.2859N) are used for a 25.0 mL sample, chlorosity = 47.23 × 0.2859 × 0.0355 × 1000/25 = 19.17‰. From Table 3, salinity = 33.82‰.

  iv. Alternatively, convert chlorosity to chlorinity by subtracting the appropriate factor given in Table 4 (see Appendix). Record chlorinity as Cl ‰.

f. Precision and accuracy

This procedure is suitable for salinities ranging from 4 to 40‰. It is accurate to between 0.05 and 0.1‰ salinity.

C. Dissolved Oxygen

Dissolved oxygen may be measured either by the Winkler titration method or by an electronic probe analyzer. The titration method generally is considered more accurate and reliable, and if electronic field instruments are used, calibrate them against the titration over a range of dissolved oxygen concentrations. Salinity corrections may be necessary with some field instruments; these can be made readily from available tables. Electronic oxygen analyzers incorporate a membrane-surfaced probe connected to a galvanic cell. Keep the sensor moist and allow 2 to 3 min for it to come to equilibrium before making a reading.

The titration method described here is a modification of the classic Winkler procedure. It can be used for dissolved oxygen concentrations ranging from 0.005 to 8 mg/L. Oxygen concentrations below 0.1 mg/L will be slightly but not significantly low.

1. Reagents

a. Manganese sulfate solution

Dissolve 480 g $MnSO_4 \cdot 4H_2O$, 400 g $MnSO_4 \cdot 2H_2O$, or 364 g $MnSO_4 \cdot H_2O$ in distilled water, filter, and dilute to 1 L. The $MnSO_4$ solution should not give a color with starch when added to an acidified potassium iodide (KI) solution.

b. Alkaline iodide solution

Dissolve 500 g NaOH in 500 mL distilled water. Dissolve 500 g KI in 500 mL distilled water. Mix the two solutions.

c. Sulfuric acid, H_2SO_4, conc

d. Starch indicator solution

Suspend 2 g soluble starch in 400 mL distilled water. Add an approximately 20% solution of NaOH, vigorously stirring until the solution becomes clear. Let stand for 1 to 2 h. Add conc HCl until the solution is *just* acid to litmus paper; then add 2 mL glacial acetic acid. Dilute to 1 L with distilled water. Discard solution when the end point is no longer a pure blue but takes on a green or brownish tint.

Alternatively use a commercially available soluble starch powder mixture.

e. Standard sodium thiosulfate solution
  i. Dissolve 2.2 g reagent grade $Na_2S_2O_3 \cdot 5H_2O$ and 0.1 g Na_2CO_3 in 1 L distilled water. Add 1 drop CS_2/L as a preservative. Prepare the thiosulfate solution in quantity and store in a dark, well-stoppered bottle below 25°C.
  ii. Iodate solution (0.01N): Dry analytical grade $KH(IO_3)_2$ at 105°C for 1 h. Cool in dessicator. Weigh out 0.3250 g and dissolve in 300 mL distilled water by warming. Cool and dilute to 1000 mL. This solution is stable indefinitely.
  iii. Calibration of thiosulfate solution: Fill a 300 mL BOD bottle with seawater. Add 1.0 mL conc H_2SO_4 and 1.0 mL alkaline iodide solution. (Add reagents below surface of water and let sample overflow to avoid trapping air bubbles in bottle.) Mix thoroughly. Add 1.0 mL $MnSO_4$ solution and mix. Withdraw 50 mL portions to titration flasks. Use one or two flasks for blank determinations; to the other flasks add 5.0 mL 0.01N $KH(IO_3)_2$ from a clean calibrated 5 mL pipet. Let iodine liberation proceed for at least 2 min but not more than 5 min. Hold solutions below 25°C and out of direct sunlight. Titrate the iodine with the thiosulfate solution. Repeat titration at least five times and use the average volume in calculating F. When V is the titer in milliliters, then

$$F = 5.00/V.$$

The factor F should not vary over time.

2. Sampling and storage

Rinse BOD bottle twice with sample being analyzed. If the sample is obtained from a reversing bottle, take a length of rubber tubing from the top to the bottom of the BOD bottle and introduce the seawater in such a way as to minimize turbulence and sample agitation. Always keep the end of the rubber tube beneath the surface of the water as the bottle is being filled. Let water overflow from the top of the BOD bottle replacing at least one bottle volume and stopper immediately. No air should remain in the bottle. Fill BOD bottles immediately or in not more than 15 min after the sample is drawn.

Store samples in the dark or in subdued light to minimize photosynthesis by any phytoplankton that may be present but analyze within 1 h. If the analysis must be delayed, add 1.0 mL manganous sulfate reagent and 1.0 mL alkaline iodide solution to the sample, restopper immediately, and mix thoroughly until the precipitate is evenly dispersed. Avoid trapping air bubbles. Hold bottles in the dark and complete analysis as soon as possible.

3. Titration

Add 1.0 mL manganous sulfate reagent and 1.0 mL alkaline iodide solution to sample. (DO NOT REPEAT this step if the pretreated sample was held in storage and titration was delayed.) Shake bottle to distribute the precipitate and let sample warm to room temperature while precipitate settles at least one-third of the way to the bottom of the bottle.

Add 1.0 mL conc H_2SO_4, restopper the bottle, and mix to dissolve the precipitate. Do not trap air in the bottle. Within an hour of acidification transfer 50 mL to a 250 mL erlenmeyer flask by means of a pipet. Avoid loss of volatile iodine. Titrate *at once* with standard thiosulfate solution until a very pale straw color remains. Add 5 mL starch indicator and continue the titration to the first disappearance of the blue color.

4. Calculations

Subtract any blank correction from the titration to obtain the corrected titration (V, mL) and calculate the oxygen content of a sample from the following formula when a 50.0 mL portion is taken from a 300 mL BOD bottle:

$$\text{mg } O_2/L = 0.1006 \times F \times V$$

D. Hydrogen Ion Concentration (pH)

pH is a measure of the hydrogen ion activity of a sample. Many reliable pH meters are available for field use. Standardize the electrodes for field use in the laboratory with standard buffers at pH 4, 7, and 10. During field operations occasionally check against a buffer with pH nearest to the observations being made. Keep electrode moistened and protected between observations. If working in marine waters, calibrate the electrodes in full strength seawater in the laboratory. Preconditioning electrodes by soaking in seawater before use may be necessary.

Field Methodology

The earlier methods of pH determination with indicator dyes largely have been supplanted by instrumental methods which are the methods of choice.

E. Turbidity

Turbidity, or light extinction, is a measure of the clarity of the water. Turbidity can vary considerably in estuaries, principally as a function of runoff, turbulence, or plankton blooms.

1. *The Secchi disk*

The Secchi disk is a weighted white and black disk lowered on a calibrated line until it just disappears from sight in the water column. It gives a measure of the turbidity of surface waters. This simple procedure is surprisingly accurate when compared with electronic instrumentation. Although it can be used only in open waters during daylight hours, it is the method of choice.

2. *Submarine photometer*

Submarine photometers, some with various filters, work on a similar principle, except that a photocell with deck readout provides the light intensity (and/or quality) at designated depths.

3. *The alpha meter*

The alpha meter has a light source and photocell in line on a rack at a known distance apart, with deck readout. The instrument indicates the turbidity at a given depth and can be useful in discriminating layered systems.

4. *Jackson candle*

Grab samples can be analyzed in the laboratory with a Jackson candle turbidimeter. This instrument measures light loss in a sample water column using the illumination from a standard candle. It is of limited use on clearer waters because the minimum detection level is 25 Jackson turbidity units.

5. *Nephelometers*

Nephelometers measure the light scattered at 90 degrees from the source and are less subject to color interferences. Plastic reference standards are available for calibration. This is the method of choice for grab samples, for it has the versatility for use on open water or effluent samples.

F. Runoff (Stream Flow, Precipitation)

Terrestrial runoff or precipitation or snow melt cover often can affect the bacteriological quality of shellfish waters. Use some index of runoff. Stream flow from many watersheds in the United States can be obtained from records at gauging stations maintained by the U.S. Geological Survey or by temporarily establishing gauging stations.

It is important to note local precipitation during the preceding 24 or 48 h on field sheets. Precipitation records are maintained by the N.O.A.A. Weather Service at many weather stations; these records are available readily.

G. Observations

"Observations" is an important category on field sheets. Record any unusual occurrences at a given site, such as atypically large flocks of waterfowl, local weather disturbances, red tides. These can be valuable in data interpretations. Additionally, note malfunctions or uncertainties in gear operation.

H. List of Suppliers*

Many instruments, such as pH meters and Jackson candles, can be obtained from most laboratory supply houses. Some instruments and gear are of such specialized nature that they are available from relatively few suppliers. The following list is not exhaustive nor complete, but is intended to be helpful:

American Optical Co. Scientific Instruments Div. Buffalo, N.Y. 14215	salinity refractometers
Beckman Instruments, Inc. 89 Commerce Road Cedar Grove, N.J. 00709	induction salinometers, DO meters

---

* This listing of suppliers does *not* constitute endorsement of the companies or of their products.

Benthos, Inc. North Falmouth, Mass. 02556	corers, general oceanographic gear
Environmental Devices Corp. Marion, Mass. 02738	bacteriological water samplers, DO meters, induction salinometers, refractometers, general oceanographic gear
General Oceanics, Inc. 5535 N.W. 7th Ave. Miami, Fla. 33127	bacteriological water samplers, reversing thermometers, general oceanographic gear
Hach Chemical Co. Box 907 Ames, Iowa 50010	nephelometers
Horizon Ecology Co. 7435 North Oak Park Ave. Chicago, Ill. 60648	bottom grabs, corers, Secchi disks
Hydrolab Corp. P.O. Box 9406 Austin, Texas 78766	DO meters, electrode salinometers
Hydro Products Co. 11777 Sorrento Valley Rd. San Diego, Calif. 92121	corers, alpha meters, general oceanographic gear
InterOcean Systems, Inc. 3510 Kurtz St. San Diego, Calif. 92110	bacteriological water samplers, corers, photometers, alpha meters, DO meters, induction salinometers, reversing thermometers, general oceanographic gear
Kahl Scientific Instrument Corp. P.O. Box 1166 El Cajon, Calif. 92022	bacteriological water samplers, bacteriological sediment samplers, bottom grabs, corers, Secchi disks, photometers, alpha meters, nephelometers, induction salinometers, salinity hydrometers, reversing thermometers, general oceanographic gear

Montedoro-Whitney P.O. Box 1401 San Luis Obispo, Calif. 93406	submarine photometers, alpha meters, DO meters, induction salinometers
Wildlife Supply Co. 301 Cass St. Saginaw, Mich. 48602	bottom grabs, corers, Secchi disks, reversing thermometers, general sampling gear
Yellow Springs Instrument Co. Yellow Springs, Ohio 45387	DO meters, electrode salinometers

2.4 REFERENCES

1. Bordner, R. & J. Winter, eds. 1978. *Microbiological Methods for Monitoring the Environment. Water and Wastes.* U.S. Environmental Protection Agency, Cincinnati, Ohio.
2. Holmes, N. A. & A. D. McIntyre. 1971. *Methods for the Study of Marine Benthos.* International Biological Programme Handbook No. 16. F. A. Davis Co., Philadelphia, Pennsylvania.
3. Sundstrom, G. T. 1975. Commercial fishing vessels and gear. U.S. Fish and Wildlife Circular 48:1-48.
4. Rounsefell, G. A. 1975. *Ecology, Utilization and Management of Marine Fisheries.* C. V. Mosby Co., St. Louis, Missouri.
5. U.S. Naval Hydrographic Office. 1970. *Instruction Manual for Obtaining Oceanographic Data.* Publication #107, 3rd ed. U.S. Government Printing Office, Washington, D.C.

CHAPTER 3

PROCEDURES FOR THE BACTERIOLOGICAL EXAMINATION OF SEAWATER AND SHELLFISH

N. Neufeld

3.1 INTRODUCTION

Three procedures for the enumeration of bacteria in seawater and shellfish are presented herein. The first two, the multiple tube fermentation technique and the membrane filter technique, are used to measure coliform and/or fecal coliform bacteria, while the third counts all the colony forming units capable of growth under the test conditions. This last test attempts to assess the number of viable, heterotrophic bacteria in the sample.

A. Multiple Tube Fermentation Technique

This technique uses the principle of dilution to extinction to estimate the number of bacteria in a sample. Decimal dilutions of the sample are introduced into replicate tubes of a medium designed to select for growth the particular organism being enumerated. Thus it reasonably can be assumed that the maximum dilution (minimum inoculum) at which growth occurs represents a volume containing a single organism. The results of such an analysis are expressed in terms of the Most Probable Number (MPN). This represents an estimate based on probability formulae. The MPN is relatively imprecise but the imprecision diminishes with an increase in the number of tubes used in each dilution. In routine analyses, five tubes in each of three decimal dilutions are used.

The method has a number of advantages, such as requiring minimal personnel training (making observations and recording results is simple); applicability to samples containing a relatively high concentration of suspended matter; and applicability to samples containing possibly toxic or inhibitory substances that tend to be neutralized by dilution. Among the method disadvantages are the limitations of sample size (inocula conventionally are limited to 10 mL per tube); the MPN table, even at the 5-tube, 3-dilution level, has an inherently poor precision and a significant bias; laboratory manipulation time per sample is relatively high; and the technique is comparatively cumbersome for field use.

Generally, the multiple tube procedure is preferable to a plating method when coliform bacteria levels are low or when other species predominate. Furthermore, a much larger sample can be analyzed than by any plating procedure.

B. Membrane Filter (MF) Technique

Because of its structure, a matrix with a multiplicity of closely packed pores of uniform size, the membrane filter quantitatively removes organisms from a liquid and retains them on or near the surface for analysis by colony culture or direct microscopic observation. Thus, the membrane filter provides a direct count of those entrapped organisms capable of growth on the culture medium provided.

The membrane filter technique is direct, reproducible, and rapid. Another advantage of using it is the flexibility provided in choosing sample size. However, particulates in the sample can limit the usable sample size. Prefiltering turbid water removes suspended matter, but it can result in a reduction of bacteria by an unpredictable 20 to 80 percent. Therefore, the membrane filter technique usually is not applicable to turbid waters. Chlorinated effluents, saline waters, or water containing high levels of domestic or industrial wastes may contain a high proportion of stressed cells of the target population. Temperature acclimation is essential for cell resuscitation. The two-step MF enrichment method improves recovery for stressed fecal coliforms from marine waters or chlorinated effluents.

An additional advantage of the membrane filter technique is that less preparation time is required so that overall costs may be less than those for the MPN method.

BACTERIOLOGICAL EXAMINATION

C. Standard Plate Count (Heterotrophic Plate Count)

This technique counts macroscopic bacterial colonies that develop after a sample is mixed with a standardized agar medium and incubated under standardized conditions. It measures colony forming units (CFU) rather than all viable organisms because under the specified conditions, not all viable organisms may produce countable colonies. The technique thus tends to underestimate the number of viable organisms.

D. Coliform Bacteria

The coliform group (also called the total coliform group) comprises all aerobic and facultative anaerobic gram-negative, nonspore-forming, cytochrome-oxidase-negative, rod-shaped bacteria that ferment lactose with gas formation within 48 h at 35°C.[1] Historically, this group has been used to demonstrate unsanitary conditions because it is associated with the intestinal contents of warm-blooded vertebrates.[2] When coliform bacteria are used as indicator organisms, their primary role is to serve as a measure of fecal contamination and thus, potentially, of the presence of enteric pathogens. As a rule they are not pathogenic and therefore do not signify directly the presence of pathogenic microorganisms. This indirect method of testing water for potential risk is relatively simple and provides an effective means of detecting conditions that identify unnecessary risk to public health.

The sanitary significance of the coliform group has been a subject of controversy in recent years because some of the genera within the group are distributed widely in nature and frequently are associated with surface runoff. The total coliform group consists of a heterogenous collection of genera, including *Escherichia, Klebsiella, Enterobacter,* and *Citrobacter. Escherichia coli* is the member of the coliform group that most closely is associated with fecal pollution. Other coliform genera—*Klebsiella, Enterobacter,* and *Citrobacter*—also may be present in feces, but usually are present in comparatively small numbers. *Klebsiella, Citrobacter,* and *Enterobacter* frequently are found in soils and on vegetation[3,4] and are capable of multiplying in polluted water.

Citrobacter and *Enterobacter* are considered to have little sanitary significance and are very common in surface runoff. *Klebsiella* is ubiquitous and is found in large numbers in waters receiving carbohydrate-rich industrial effluents such as those from pulp mills, textile mills, and sugar refineries.[5]

E. Fecal Coliform Group

The inadequacy of the total coliform test in differentiating fecal from nonfecal pollution led to the development of the fecal coliform test. In terms of the MPN test fecal coliform bacteria are defined as coliforms that have the ability to ferment lactose in EC medium with gas production at 44.5°C within 24 h. They also are referred to as thermotolerant coliforms. They are considered to be more directly associated with fecal contamination from warm-blooded vertebrates than are other members of the coliform group. Because of this relationship between fecal coliforms and fecal pollution, the National Shellfish Sanitation Program has accepted the group as an indicator of fecal pollution

It should be emphasized that the term "fecal coliform," like the term "coliform," has no taxonomic validity. Therefore, the meaning of a fecal coliform count becomes clear only when it is expressed in terms of the test procedure. A fecal coliform test estimates *E. coli* densities, but the proportion of the other fecal coliforms present will vary, depending on the sample source.[6] In shellfish-growing waters, *E. coli* usually makes up a high proportion of the fecal coliform count—75% to 95%,[7,8] but in other waters, particularly those receiving effluents rich in carbohydrates, the test is much less specific for *E. coli*. In such waters the incidence of *Klebsiella* is markedly increased. Some *Klebsiella* isolated from certain industrial effluents that are demonstrably free from fecal contamination, nevertheless, will be enumerated as fecal coliform organisms.[9]

This occasional lack of specificity in the fecal coliform group is one of the arguments put forward for preferring *E. coli* as the fecal indicator. The justifications for this proposal are that the natural habitat for *E. coli* is the lower intestinal tract of vertebrates, that it does not multiply readily in water, and that it is more abundant in feces than are other coliforms or pathogenic bacteria.

3.2 MULTIPLE TUBE FERMENTATION TEST FOR COLIFORM AND FECAL COLIFORM BACTERIA

A. Sample Collection and Handling

See Section 2.2, Collection of Bacteriological Samples.

Whenever possible, initiate the bacteriological examination immediately after sample collection, preferably within 1 h. When conditions necessitate delay, keep sample at or below 10°C, preferably at 4°C, until examined. In no case examine shellfish samples that have been held longer than 24 h or seawater samples held longer than 30 h.

B. Lauryl Sulfate Tryptose (LST) Fermentation Test

This test involves at least two steps, a presumptive test that provides presumptive evidence of the presence of total coliform and fecal coliform bacteria and a confirmed test that, depending on the medium and incubation temperature used, confirms the presence either of total coliform or fecal coliform bacteria (see Figure 3:1).

The official methods of most water pollution and shellfish control agencies include the presumptive test followed by the confirmed test for fecal coliform bacteria (LST-EC method).[1,10,11]

1. *Presumptive test*

The presumptive test consists of incubating selected volumes of sample in LST broth for a maximum of 48 ± 3 h in an air incubator at 35 ± 0.5°C. When a positive tube is found (as indicated by the presence of gas), a portion of the culture is transferred to EC broth for the confirmed fecal coliform test and, if applicable or desired, a separate transfer is made to BGB broth for the confirmed total coliform test. The procedure requires examination of the culture tube at 24 ± 2 h incubation. Any LST tubes still negative at the 48 h reading are considered negative for the entire coliform group.

    a. Inoculation procedure

        Use LST broth in fermentation tubes in the following strengths: double strength (10 mL per 150 mm × 20 mm tube), 1.5 strength (20 mL per 150 mm × 25 mm tube), or single strength (10 mL per 150 mm × 15 mm tube). When

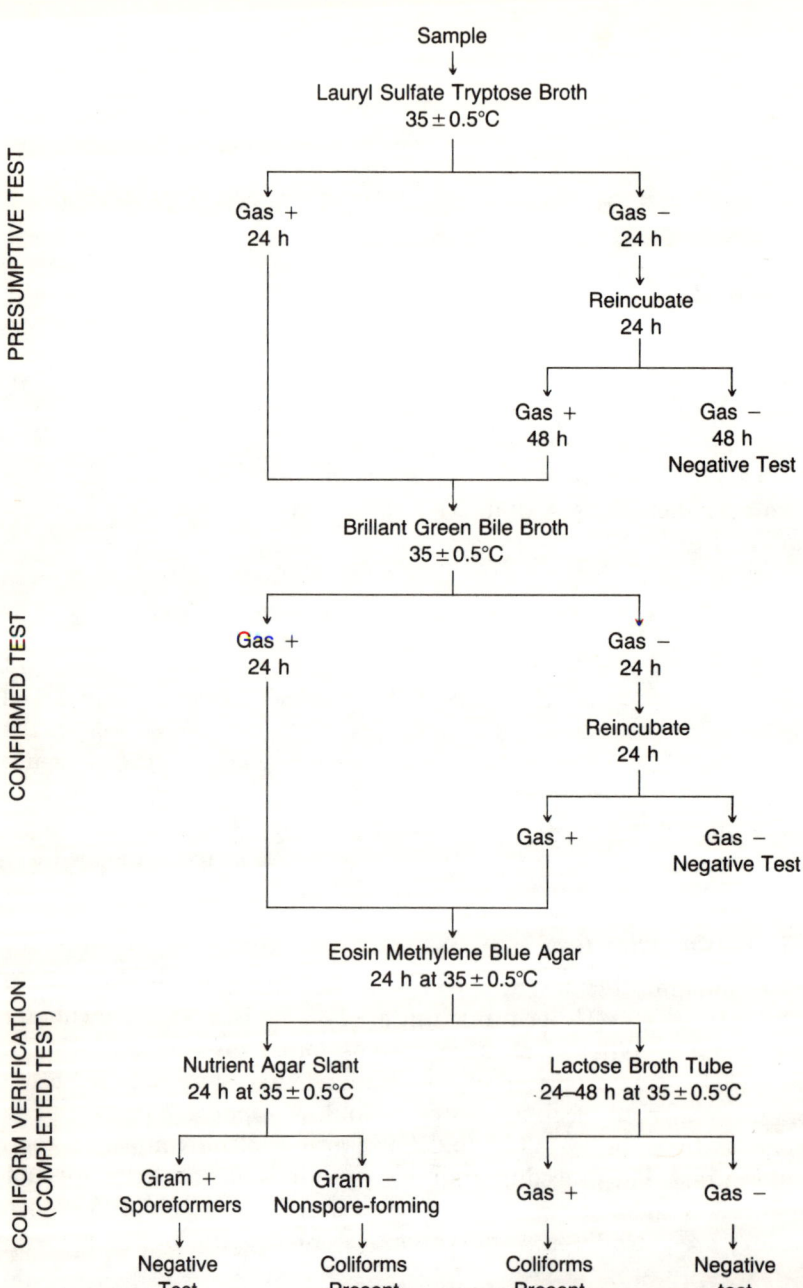

Figure 3:1—Flow chart for the total coliform MPN test.

inoculating volumes greater than 1 mL, select the medium strength so that the added sample will not dilute the medium below recommended use-strength. Possible inhibitory effects of sample constituents that might be introduced with large inocula (10 mL or more) will be reduced if the larger medium volume is chosen (20 mL of 1.5 strength).

Test five replicates in each of three volumes. Select volumes that are log multiples of 10 mL (decimal dilutions). Most shellfish-growing water samples will yield determinate results when five 10 mL, five 1 mL, and five 0.1 mL portions of sample are tested. For routine monitoring of waters of known quality, three replicates per dilution may be used. Make no transfer of less than 1 mL; thus dispense 1 mL of a 1:10 dilution rather than 0.1 mL of the undiluted sample. Set up, preferably in a rack, and label the requisite fermentation tubes containing LST broth in the desired strengths.

Before transferring the sample to the culture tube or dilution bottle or tube, shake sample bottle vigorously 25 complete up-and-down movements of about 0.3 m in 7 s. Using approved pipets (see Chapter 1) transfer the desired sample volumes. Do not use any pipet to deliver less than one-tenth of its total volume. Inoculate five 10 mL and five 1 mL volumes of sample and five 1 mL volumes of a decimal dilution of the sample.

b. Incubation

Incubate inoculated tubes at $35 \pm 0.5°C$. After 24 ± 2 h examine tubes for gas production, as evidenced by a bubble of gas in the inverted vial or by effervescence when the tube is shaken gently. Remove gas-positive tubes and reincubate gas-negative ones. Examine reincubated tubes after 48 ± 3 h. All tubes not showing gas at this time are negative for the entire coliform group and may be discarded.

Submit gas-positive tubes to the confirmed test as soon as possible after they are found.

2. *Confirmed test*

After incubating in LST broth, transfer from all gas-positive tubes to BGB broth and/or to EC broth (see Figure 3:2). To be considered

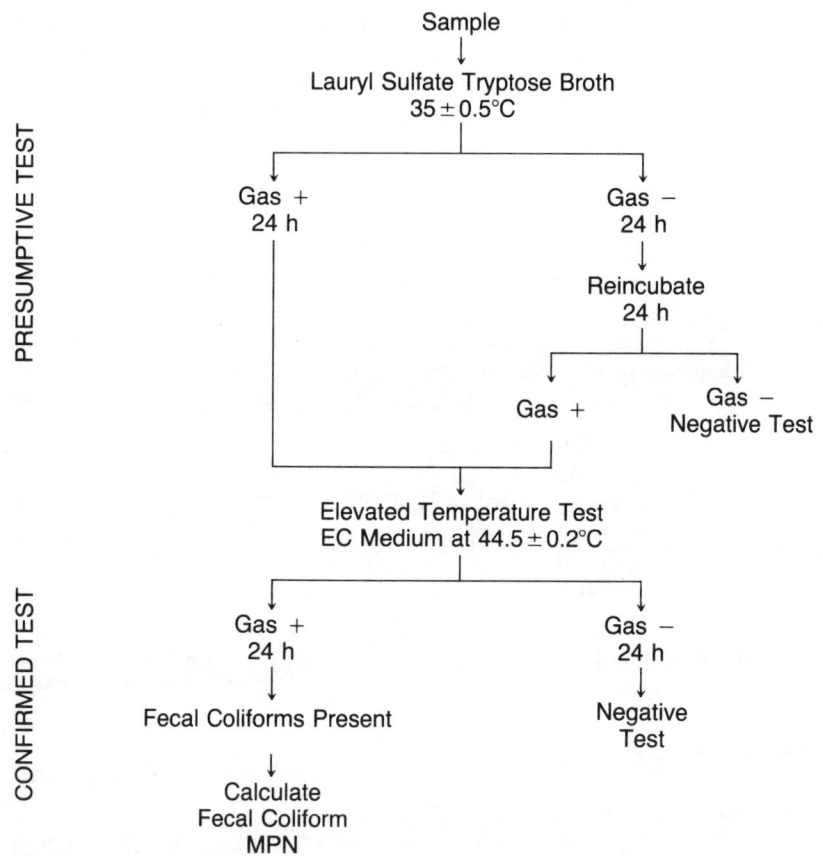

Figure 3:2—Flow chart for fecal coliform MPN test.

positive for the more inclusive total coliform group, the organisms must grow and produce gas within 48 ± 3 h at $35 \pm 0.5°C$ when inoculated into BGB broth. To be considered positive for the fecal coliform group, the organisms must grow and produce gas in EC medium, at $44.5 \pm 0.2°C$ within 24 ± 2 h. Confirmed tests for both groups may be run from a single gas-positive LST tube.

    a. Inoculation and incubation

        Carefully agitate each positive LST tube by swirling. Using a bacterial transfer loop (3 mm diam or larger) or a sterile hardwood applicator stick, transfer growth from the positive tube to the confirming tube. If both BGB and EC broths are to be inoculated, inoculate the EC broth first to avoid transferring inhibitory dyes from the BGB tube.

Promptly place inoculated EC tubes in a water bath held at $44.5 \pm 0.2°C$. Adjust bath water level to at least 1 cm above the level of the medium in the tubes. Incubate BGB tubes in an air incubator at $35 \pm 0.5°C$.

b. Reading results

After 24 ± 2 h incubation examine EC tubes for growth and gas. Gas-positive tubes are positive for the fecal coliform group. Tubes without gas are considered negative for the group even if growth is present.

Examine BGB tubes at 24 ± 2 h and record results. The presence of gas is a positive test for the total coliform group. Reincubate all negative tubes an additional 24 h, and reexamine. Record results and discard tubes.

c. Computing and recording results

Record the number of gas-positive tubes in each dilution. From the appropriate MPN Table (Tables 3:1 and 3:2), determine the MPN value for the test. The values shown in the tables are for inocula of 10 mL, 1 mL, and 0.1 mL; with Table 3:1 for five tubes per dilution and Table 3:2 with three tubes per dilution. Calculate values for larger or smaller sample portions by multiplying table figures by the appropriate dilution factor (see Table 3:3). Use only three dilutions in the code for calculating an MPN value. To obtain the proper code, choose the smallest sample volume giving all positive results and the two consecutive (higher) dilutions. If four consecutive dilutions produce some positive results, add the number of positive tubes in the highest dilution to those in the next lower dilution.

C. A-1-M Fermentation Test for Fecal Coliform Bacteria

The modified A-1 method (A-1-M) provides fecal coliform results that are equivalent to those produced by the LST-EC method.[12] It

TABLE 3:1. MPN INDEX* AND 95% CONFIDENCE LIMITS

Result			MPN	Category**		95% Confidence Limits	
				1	2	Lower	Upper
0	0	0	<2				
0	0	1	2		x	<0.5	7
0	1	0	2	x		<0.5	7
0	2	0	4		x	<0.5	11
1	0	0	2	x		<0.5	7
1	0	1	4		x	<0.5	11
1	1	0	4	x		<0.5	11
1	1	1	6		x	<0.5	15
1	2	0	6		x	<0.5	15
2	0	0	5	x		<0.5	13
2	0	1	7		x	1	17
2	1	0	7	x		1	17
2	1	1	9		x	2	21
2	2	0	9	x		2	21
2	3	0	12		x	3	28
3	0	0	8	x		1	19
3	0	1	11	x		2	25
3	1	0	11	x		2	25
3	1	1	14		x	4	34
3	2	0	14	x		4	34
3	2	1	17		x	5	46
3	3	0	17		x	5	46
4	0	0	13	x		3	31
4	0	1	17	x		5	46
4	1	0	17	x		5	46
4	1	1	21		x	7	63
4	1	2	26		x	9	78
4	2	0	22	x		7	67
4	2	1	26		x	9	78
4	3	0	27	x		9	80
4	3	1	33		x	11	93
4	4	0	34		x	12	93
5	0	0	23	x		7	70
5	0	1	31	x		11	89
5	0	2	43		x	15	110
5	1	0	33	x		11	93
5	1	1	46	x		16	120
5	1	2	63		x	21	150
5	2	0	49	x		17	130
5	2	1	70	x		23	170
5	2	2	94		x	28	220

TABLE 3:1. (continued)

Result			MPN	Category**		95% Confidence Limits	
				1	2	Lower	Upper
5	3	0	79	x		25	190
5	3	1	110	x		31	250
5	3	2	140	x		37	340
5	3	3	180		x	44	500
5	4	0	130	x		35	300
5	4	1	170	x		43	490
5	4	2	220	x		57	700
5	4	3	280		x	90	850
5	4	4	350		x	120	1000
5	5	0	240	x		68	750
5	5	1	350	x		120	1000
5	5	2	540	x		180	1400
5	5	3	920	x		300	3200
5	5	4	1600	x		640	5800
5	5	5	≥2400			800	—

* MPN/100 mL (or /100 g) using sample portions of 5 × 10 mL, 5 × 1 mL, and 5 × 0.1 mL (or g).

** Category 1: Normal results, obtained in 95% of cases.

Category 2: Less likely results, obtained in only 4% of cases. Results even less likely to occur have been omitted from the table. If they occur in more than 1% of the tests, it is an indication of faulty technique or that the assumptions underlying the MPN index are not being fulfilled.

has the advantages of requiring only a single medium and no culture transfers, and it gives a final result in 24 h.

1. Procedure

Follow the procedures of sampling and inoculation given in Section 3.2B, substituting A-1 medium for LST broth.

Incubate inoculated tubes at $35 \pm 0.5°C$, preferably in a water bath, for 3 ± 0.5 h. Transfer tubes to a water bath at $44.5 \pm 0.2°C$ and incubate for an additional 21 ± 2 h.

2. Reading, computing, and recording results

A positive test is indicated by a bubble of gas in the inverted vial or by active effervescence when the tube is shaken gently. Compute and record results as directed in Section 3.2B.

TABLE 3:2. MPN INDEX* AND 95% CONFIDENCE LIMITS

Result	MPN	Category** 1	Category** 2	95% Confidence Limits Lower	95% Confidence Limits Upper
0 0 0	<3				
0 1 0	3		x	<0.5	13
1 0 0	4	x		<0.5	20
1 0 1	7		x	1	21
1 1 0	7	x		1	23
1 2 0	11		x	3	36
2 0 0	9	x		1	36
2 0 1	14		x	3	37
2 1 0	15	x		3	44
2 1 1	20		x	7	89
2 2 0	21	x		4	47
3 0 0	23	x		4	120
3 0 1	39	x		7	130
3 1 0	43	x		7	210
3 1 1	75	x		14	230
3 2 0	93	x		15	380
3 2 1	150	x		30	440
3 2 2	210		x	35	470
3 3 0	240	x		36	1300
3 3 1	460	x		71	2400
3 3 2	1100	x		150	4800
3 3 3	≥2400				

* MPN/100 mL (or /100 g) using sample portions of 3 × 10 mL, 3 × 1 mL, and 3 × 0.1 mL (or g).

** Category 1: Normal results, obtained in 95% of cases.

Category 2: Less likely results, obtained in only 4% of cases. Results even less likely to occur have been omitted from the table. If they occur in more than 1% of the tests, it is an indication of faulty technique or that the assumptions underlying the MPN index are not being fulfilled.

D. Examination of Shellfish

The multiple tube fermentation techniques given above are applicable to the examination of shellfish as well as seawater. Sample preparation obviously is different. The A-1-M procedure for fecal coliforms is not recommended for the routine examination of shellfish.

BACTERIOLOGICAL EXAMINATION

TABLE 3:3. ADJUSTING MPN TO ALTERNATIVE SAMPLE VOLUMES

Inoculum Size, mL					Code and	
10	1	0.1	0.01		Calculations	MPN/100 mL
5/5*	5/5	0/5	0/5	=	5:0:0 × 10	230
5/5	5/5	2/5	0/5	=	5:2:0 × 10	490
5/5	4/5	2/5	0/5	=	5:4:2	220
0/5	1/5	0/5	0/5	=	0:1:0	2
5/5	3/5	1/5**	1/5	=	5:3:2	140
5/5	5/5	3/5	1/5	=	5:3:1 × 10	1100

* No. positive/No. tubes inoculated.
** Number adjusted by adding positive tube(s) from highest dilution to next lower.

1. **Sample collection**

   See Section 2.2, Collection of Bacteriological Samples.

2. **Sample preparation**

   a. Shellfish in the shell

   i. Shell cleaning

   First scrub the analyst's hands thoroughly with soap and water. If heavy rubber or, stainless steel, or otherwise fabricated gloves are worn as hand protection, scrub these with soap and water. If sterile cotton gloves are worn, use a fresh pair for each sample. Discard shellfish with badly broken shells or those that are dead as evidenced by permanently gaping shells.

   Scrape extraneous material from the shell and using a sterile brush scrub the shellstock under potable running water paying particular attention to crevices at shell junctions. Place cleaned shellstock in a clean container or on clean towels and let air dry.

   ii. Removal of shell contents

   Before starting to remove shell contents, thoroughly scrub the hands (or the protective gloves if they are not sterile) with soap and water, rinse with water, and rinse again with 70% alcohol. Alternatively, protective gloves may be dipped into an iodophor solution and rinsed under running water of potable quality. Open the shellfish as directed below, collecting shell liquor and meats in a sterile blender jar or other suitable sterile container.

Oysters

Hold the oyster in the hand or on a fresh clean paper towel on the bench with the deep shell on the bottom and the hinge pointed away from the analyst. Use a sterile oyster knife and insert the point between the shells. Cut the adductor muscle from the upper flat shell and pry the shell open wide enough to drain the shell liquor into a sterile tared beaker, widemouthed jar, or blender jar. Pry the upper shell loose at the hinge, discard it and transfer the meats to the container after severing the muscle attachment to the lower shell.

Hard Clams

Entry into the hard clam, *Mercenaria mercenaria*, or other species with tightly fitting shells is done best with a sterile, thin-bladed knife similar to a paring knife. To open the clam, hold it in the hand, place the edge of the knife at the junction of the bills, and force the knife between the shells with a squeezing motion. Alternatively, cut a small hole in the bill with sterile bone-cutting forceps, insert the knife, and sever the two adductor muscles. Drain shell liquor into sample container. Cut adductor muscles from the shells and transfer the body of the animal to the container.

Other Clams

The soft clam, *Mya arenaria*, the Pacific butter clam, *Saxidomus giganteus*, the surf clam, *Spisula solidissima*, and similar species may be shucked with a sterile paring or clam knife or with a sharp flexible spatula, by entering at the siphon and cutting the adductor muscles first from the top valve and then from the bottom valve.

Mussels

Volsella and *Mytilus* species may be entered at the byssal opening. Remove the byssal threads during cleaning of the shell. Insert the knife and spread the shells apart with a twisting motion, letting the liquor drain. Cut away the many attachments from the shell.

BACTERIOLOGICAL EXAMINATION

b. Shucked shellfish

Transfer a suitable quantity from a sample jar to an appropriate container with a sterile spoon.

3. Sample blending

Use the following procedure whenever feasible: Shuck 10 to 12 shellfish into a tared sterile blender jar or beaker and add an equal weight of diluent (phosphate buffer or 0.1% peptone). Blend for 60 to 90 s and dilute 1:10 by promptly adding 20 g of homogenate to 80 mL of diluent. Prompt transfer assures that the blended sample does not separate out in the blender jar. A widemouthed 25-mL pipet is convenient but a smaller one, e.g., 10-mL, may be used. When the shucked quantity from 10 specimens greatly exceeds 200 g and when sample consistency permits, grind undiluted for 30 s, transfer 200 g of this preliminary grind to a second sterile blender jar, add an equal weight of diluent, and proceed as above.

When the consistency of a 1:2 dilution is too thick for effective blending, use 100 g of shucked meats and add 300 mL of diluent. Blend for 120 s and transfer 40 g of the ground material to 60 mL of diluent. This produces a 1:10 dilution suitable for analysis.

When 10 shellfish yield a quantity of shucked material much less than 200 g, make a 1:10 dilution directly in the blender jar by adding 90 mL diluent for every 10 g sample. Blend for 90 s. When specimens are very large, use a large blender jar or several blenders and pool the homogenate in a large sterile beaker or jar. Alternatively, grind the mollusk in a sterile food chopper, mix the grindings thoroughly with sterile implements, and homogenize a 200 g portion.

To test edible portions only, remove sections of those portions to give an appropriate weight for blending.

4. Sample dilution

Proceed with the examination within 2 min of blending. If examination is delayed beyond 2 min, resuspend the sample immediately before making further dilutions, or before inoculating into the culture medium. This will avoid gross errors due to the separation of the ground sample on standing. The multiple tube procedure, when applied to shellfish, introduces a factor not encountered in the analysis of seawater; the largest sample portion routinely inoculated is 10 mL of the 1:10 dilution, representing 1 g of shellfish meats. The large infusion of organic matter may affect the selective properties of the medium and produce false-positive results. Thus, it may be preferable to add 10 mL of inoculum to tubes containing 20 mL of 1.5 strength medium rather than to 10 mL of double strength or 5 mL triple strength medium.

Follow the procedure given in Section 3.2B. Note however that the largest inoculum routinely will be 1 g of sample (10 mL of the 1:10 dilution) which will necessitate an adjustment in MPN computation.

Continue preparation of dilutions by transferring 10 mL of the initial 1:10 dilution to 90 mL of diluent (1:100). Transfer 10 mL of the 1:100 dilution to 90 mL of diluent to obtain a 1:1000 dilution. To obtain the desired inoculum levels of 1 g, 0.1 g, 0.01 g, and 0.001 g, transfer 10 mL and 1 mL portions of the 1:10 dilution, 1 mL of the 1:100 dilution, and 1 mL of the 1:1000 dilution, respectively, into the test medium. Alternatively, prepare 1:10 dilutions by adding 11 mL of sample portion to 99 mL diluent.

5. Modified MacConkey procedure for fecal coliforms

This plate count procedure may be used to determine fecal coliform bacteria in both hard and soft clams. Prepare samples and dilutions as indicated in Section 3.2D, 1–4. Add 6 g of homogenate to a 180 mL bottle containing 54 mL sterile phosphate-buffered saline (PBS). Mix well and add 60 mL melted (45 to 50°C) double strength modified MacConkey agar to the bottle and mix gently. Distribute the mixture into 10 previously labelled petri dishes (100 × 15 mm). Let agar solidify, invert plates, and incubate at 45.5 ± 0.5°C for 24 h. Count appropriate colonies using a Quebec or similar type counter.

Fecal coliforms appear as brick red, subsurface, elliptical colonies greater than 0.5 mm in diameter. Count only those colonies having these characteristics. Sum the fecal coliform colonies appearing on all 10 petri dishes, multiply by the appropriate dilution factors, and report results as "number of fecal coliforms/100 g of sample."

Prepare PBS by mixing 8.5 g sodium chloride, NaCl, and 1.25 mL stock phosphate buffer solution and diluting to 1 L with distilled water.

3.3 DIFFERENTIATION OF COLIFORM ORGANISMS

In conducting special studies to determine source or kind of pollution, to evaluate methods, or for other purposes, it may be desirable to differentiate among coliform types. This procedure is based on the IMViC test series (*I*ndole production, *M*ethyl red, *V*oges-Proskauer test, and *C*itrate utilization). The procedures may be applied to any cultures of coliform bacteria, independent of origin.

A. Coliform Verification

The following procedure had been identified previously as the Completed Test for coliform bacteria.

Streak from any coliform-positive tube or colony to a plate of Levine's EMB agar, invert the plate, and incubate at $35 \pm 0.5°C$ for 24 ± 2 h. Typical coliform colonies are discrete and nucleated with or without a metallic sheen. Colored colonies that may be coalescent and mucoid, with a weak sheen, may be coliforms. Colorless colonies are unlikely to be lactose fermenters and, therefore, are not coliforms.

From each EMB plate pick one or more typical coliform colonies. If no typical colonies are present, pick two or more colonies most likely to be coliforms. From each colony picked inoculate a tube of lactose broth (or EC broth if fecal coliforms are sought) and a nutrient agar slant. Incubate at $35 \pm 0.5°C$ for 24 ± 2 h (incubate EC broth tubes in a water bath at $44.5 \pm 0.2°C$ for 24 ± 2 h). Record gas formation; if no gas is formed in lactose broth incubate an additional 24 ± 2 h. If no gas is formed in 48 h the colony was not a coliform (no gas in 24 h in EC broth is a negative result for fecal coliforms). From the agar slant prepare a smear, stain by the gram procedure, and examine microscopically.

If gas is produced in the fermentation tube and gram-negative nonspore-forming rods are observed, the culture is a coliform; proceed with differential tests.

B. Cytochrome Oxidase Test

With a platinum needle transfer a small amount of growth from the nutrient agar slant to a filter paper impregnated with a 1%

solution of tetramethyl-*p*-phenylamine diamine dihydrochloride. A dark color developing within 30 s constitutes a positive test. Disregard color changes that take place more slowly. Do not use cultures older than 24 h for this test. Coliform organisms are oxidase-negative. Commercial test papers for this determination are available.

C. IMViC Tests

Perform the following tests on oxidase-negative coliform isolates. Inoculate, with a needle, a Simmons citrate agar slant and a fermentation tube of lactose broth with pH indicator. Use a minimal inoculum to prevent carry-over of nutrients that might lead to false-positive results. Inoculate, with a loop, a tube of MR-VP medium and a tube of 1% tryptone. Incubate all tubes at $35 \pm 0.5°C$.

1. Indole test

After 24 h add 3 to 4 drops Kovac reagent to the tryptone tube. A red color developing within 2 min in the floating reagent layer is a positive reaction.

2. Lactose fermentation

Read the lactose tube at 24 and 48 h. Acid and gas development is a positive reaction.

3. Voges-Proskauer reaction

After 48 h transfer 1 mL of the growing culture in the MR-VP tube to a clean tube and add 0.6 mL alpha-naphthol solution and 0.2 mL 40% KOH solution. Shake vigorously and let stand for 2 to 4 h. A pink to red color is a positive reaction.

4. Citrate utilization

Examine the Simmons citrate slant after 48 h. Growth on the slant with development of a blue color is a positive reaction.

5. Methyl red reaction

After 5 days of incubation add to the MR-VP tube 5 drops methyl red indicator. A red color is positive; a yellow color is negative.

6. Proposed alternative MR-VP procedures

When the proposed alternative MR-VP procedure is to be used, make parallel evaluations using the procedure given above to establish the relationship between the methods.

Inoculate a brain heart infusion agar slant. Incubate at $35 \pm 0.5°C$ for 4 to 6 h (until visible growth occurs). From the slant, heavily inoculate two tubes containing 0.5 mL sterile MR-VP medium. Incubate at $35 \pm 0.5°C$ for 18 to 24 h.

a. For the MR test add one drop methyl red indicator to one tube. A red color is positive; a yellow color is negative.

b. For the VP test add to the second tube, in sequence, 2 drops 0.5% creatine solution, 3 drops alpha-naphthol solution, and 2 drops KOH solution. Agitate vigorously after each addition. Examine after 30 min; a pink or red color is positive.

7. Interpretation of IMViC reactions

Organism	Indole	Methyl Red	Voges-Proskauer	Citrate
Escherichia coli	+ or −	+	−	−
Citrobacter freundii	−	+	−	+
Klebsiella/Enterobacter	− or (+)	−	+	+

3.4 MEMBRANE FILTER (MF) METHOD FOR FECAL COLIFORM (AND COLIFORM) BACTERIA

When the MF method is to be introduced into a laboratory, it is important that parallel evaluations using the MPN procedure be carried out to establish the relationship between the two methods. Make parallel tests seasonally to evaluate different shellfish growing areas. (The two-step procedure for total coliform bacteria enhances recovery of stressed organisms.)

A. Sample Collection and Handling

See Section 2.2, Collection of Bacteriological Samples.

B. Inoculation Procedure

For maximum accuracy, filter samples of such volume to yield counts between 20 and 60 fecal coliform colonies (or 20 and 80 total coliform colonies). Number and label all petri dishes, graduated cylinders, 30 mL blanks, and 90 mL (or 99 mL) dilution blanks with the proper laboratory reference number. Mix sample thoroughly by shaking in the prescribed manner (see Section 3.2B.3) and distribute chosen sample volumes into sterile graduated cylinders or dilution blanks. Prepare necessary dilutions and mix as above.

For fecal coliform testing place a sterile absorbent pad in each labelled culture dish and pipet approximately 2 mL M-FC medium to saturate the pad. Carefully remove any surplus liquid from the culture dish. Alternatively, pipet 8 to 10 mL sterile melted M-FC agar into the culture dish. (For coliform testing use 8 to 10 mL M-Endo agar LES, place an absorbent pad in the cover of the dish, and saturate the pad with 1.8 to 2.0 mL LST broth.)

Using forceps sterilized by burning off alcohol that adheres after the forceps are dipped in it, place a sterile filter membrane over the support screen in the filtering apparatus, grid side up. Carefully place the matched funnel unit over the receptacle and lock into place. Pass sample through the filter under partial vacuum. Rinse filter by filtering three 20 to 30 mL volumes of sterile diluent (buffer or peptone) between samples. Unlock and remove funnel, immediately remove filter membrane with sterile forceps, and place it on the prepared sterile pad or agar medium with a rolling motion to ensure intimate contact with the medium and to avoid air entrapment. For initial incubation of coliforms place membrane on surface of pad saturated with LST broth in petri dish cover. Do not directly filter sample volumes less than 30 mL. Add such smaller volumes to a 30 mL dilution blank, agitate in the prescribed manner, and filter the total contents of the dilution blank.

C. Incubation

Incubate the inoculated culture dishes at $35 \pm 0.5°C$ for 2 h. Place the plates in waterproof plastic bags, and submerse them, inverted, in a water bath at $44.5 \pm 0.2°C$ for 24 h. Anchor dishes below the water surface in the bath to maintain the critical temperature requirements. Alternatively, use a heat-sink incubator at $44.5 \pm 0.2°C$.

(For coliforms, after incubating for 1.5 to 2.5 h at $35 \pm 0.5°C$, transfer membrane from absorbent pad to surface of agar medium in the same dish, leave pad in place, invert dish, and incubate at $35 \pm 0.5°C$ for 22 ± 2 h.)

Run a blank filter membrane at the beginning of a sample series to test funnel assembly and diluent sterility. Repeat the blank filtering at the end of the sample series.

Complete the analysis from first dilution through filtration to placement of the filters on the medium in 20 min.

D. Counting

Count colonies with the aid of a low-power (10 to 15×) binocular wide-field dissecting or stereoscopic microscope, using fluorescent light to provide illumination approximately perpendicular to the membrane surface. Colonies produced by fecal coliforms are blue. Nonfecal coliform bacteria produce gray or cream-colored colonies. Background on the membrane will vary from yellowish-cream to faint blue, depending on the age of the rosolic acid salt reagent and the chemistry of the membrane filter. Normally, few nonfecal coliform colonies will develop on M-FC medium because of the selective action of the elevated temperature. (Coliform colonies on M-Endo agar LES are red with a metallic sheen.)

E. Computing and Reporting Results

Compute the density from the sample volumes that produce MF counts within the desired colony range. Report the calculated count as total or fecal coliforms per 100 mL depending on the medium used.

The computation is derived from the count on the membranes that falls within the desired range:

$$\frac{\text{Fecal Coliforms/100 mL}}{\text{(or total coliforms)}} = \frac{\text{colonies counted} \times 100}{\text{mL sample filtered}}$$

If the MF counts are individually less than 20, total all such counts and base the value on the total volume of sample filtered. For example, if duplicate 50 mL portions produced counts of five and three colonies, report the count as 8/100 mL. Round off all counts to two significant digits.

3.5 STANDARD PLATE COUNT (HETEROTROPHIC PLATE COUNT)

A. Sample Collection and Handling

See Section 2.2, Collection of Bacteriological Samples.

B. Inoculation Procedure

Prepare plates for counting in a clean work area free from dust and drafts and out of direct sunlight. Label and set out required dilution blanks and petri dishes.

To obtain maximum precision and accuracy select dilutions to provide counts of 30 to 300 colonies per plate. Inoculate all dilutions in duplicate. Do not measure volumes of less than 1 mL directly, rather make a dilution of the original sample and plate at least 1 mL. It is not desirable to plate more than 0.1 g of shellfish per plate because of excessive turbidity produced by larger quantities of sample. When the total number of colonies resulting from plating 0.1 g is less than 30, record the result as observed. Use dilutions prepared to test for coliform and fecal coliform organisms for the plate count.

Melt an appropriate amount of sterile plate count agar (PCA), 12 to 15 mL for each plate to be poured, in a boiling water bath or in flowing steam. Do not expose agar to heat any longer than necessary to ensure complete melting. After melting, transfer agar to a water bath at $45 \pm 1°C$. Do not use the agar until its temperature is between 44 and 46°C. Do not hold melted agar at this temperature longer than 3 h. Do not reuse agar after once melting.

Process samples in batch sizes small enough to permit all steps of the procedure, from sample dilution to pouring of the agar, to be completed in 20 min. For shellfish samples prepare the original 1:10 dilution of blended shellfish as outlined in Section 3.2D. Make further decimal dilutions as needed by transferring 10 mL of this 1:10 dilution to a 90 mL dilution blank (or 11 mL to a 99 mL blank), and continue until all desired dilutions have been prepared. Thoroughly agitate each dilution bottle by shaking for 15 s on a shaking machine, or vigorously by hand for 25 complete up-and-down movements of about 0.3 m in 7 s.

For routine work, three dilutions usually are plated, i.e., 1:10, 1:100, and 1:1000. Withdraw the desired diluted volume and deposit

BACTERIOLOGICAL EXAMINATION

it in the appropriate petri dish with the tip of the pipet touching the inside bottom of the petri dish. After the liquid has drained to the rest point in the tip of the pipet, touch tip once against a dry spot on the dish. Pour 12 to 15 mL melted, tempered agar medium into each inoculated petri dish. As soon as the agar has been added, mix sample and medium thoroughly by swirling and tilting. Use care to avoid splashing agar on dish side or lid.

C. Incubation

When the agar has solidified, invert plates and incubate at $35 \pm 0.5°C$ for 48 ± 3 h. Humidity control of the incubator may be necessary to prevent excessive moisture loss from the plates during incubation.

Prepare control plates to check on sterility of the diluent, glassware, and agar medium. The control plate also will detect possible air contamination. If a control plate indicates contamination, report all affected plates as Laboratory Accidents.

D. Counting Plates and Computing Results

Count plates immediately after incubation is completed. Use a Quebec Colony Counter or equivalent instrument.

To compute the plate count, multiply the average number on the duplicate plates of the same dilution by the reciprocal of the dilution used. Record the result as the Standard Plate Count (SPC) per g (or per mL for seawater samples) at 35°C. Avoid fictitious precision and accuracy by recording only the two left-hand digits, rounding off numbers if necessary, e.g.: report 142 as 140, and 146 as 150. Record a two-digit count, e.g., 35, as such.

Count all colonies on spreader-free plates with 30 to 300 colonies, including colonies of pinpoint size; record dilution used and report total colonies as Standard (or Heterotrophic) Plate Count per gram (Table 3:4, Sample No. 1011). If only one plate of a given dilution contains 30 to 300 colonies, count both plates, unless there are spreader colonies, and include such counts in the average. (Table 3:4, Sample No. 1012). If plates from consecutive decimal dilutions yield 30 to 300 colonies each, compute estimated count per gram for each dilution by multiplying colonies per plate by dilution used (Table 3:4, Sample No. 1013). Report arithmetic average as Standard Plate

TABLE 3:4. COMPUTING THE PLATE COUNT

Sample No.	Colonies per Dilution		Count Ratio*	Plate Count at 35°C
	1:100	1:1000		
1011	175	16		19 000
	208	17		
1012	322	23		
	278	29		30 000
1013	296	40		
	378	24		33 000
1014	138	42		
	162	30	2.4	15 000
1015	274	35		
	230	Spreader**	1.4	30 000
1016	325	22		
	329	25		33 000
1017	18	0		
	21	0		<3000

* Ratio of the greater to the lesser plate count (applied to plates from consecutive dilutions having between 30 and 300 colonies).

** Spreader and adjoining area of repressed growth covering more than ½ of the plate.

Count per gram unless the higher computed count is more than twice the lower one, in which case report lower computed count as Standard Plate Count per gram (Table 3:4, Samples No. 1014 and 1015).

If spreaders occur on plate(s) selected, count colonies on representative portions thereof only when colonies are well distributed in spreader-free areas. If area covered by the spreader(s), including total repressed-growth area, is one-quarter of total area or more or if all plates from sample have excessive spreader growth, are known to be contaminated, or are otherwise unsatisfactory, report result as spreader (Spr), Laboratory Accident (LA), or Growth Inhibitors (GI) or use other appropriate term.

Where there is no plate with 30 to 300 colonies, and one or more plates have more than 300 colonies, use such plates(s) having a count nearest 300 colonies (Table 3:4, Sample No. 1016). If plates from all dilutions yield less than 30 colonies each, record actual number of colonies in the lowest dilution unless there were spreaders and report computed count as less than 30 (× the corresponding dilution) (Table 3:4, Sample No. 1017). If colonies per plate appreciably exceed 300 in the highest dilution plates, report the results as more than 300 (× the dilution factor).

3.6 REFERENCES

1. American Public Health Association. 1981. Standard Methods for the Examination of Water and Wastewater, 15th ed., APHA, Washington, D.C.
2. Escherich, T. 1885. Die Darmbakterien des Neugeborenen und Säuglings. *Fortschr. der. Med.* 3:515.
3. Geldreich, E. E., C. B. Huff, R. H. Bordner, P. W. Kabler & H. F. Clark. 1982. The faecal coli-aerogenes flora of soils from various geographical areas. *J. Appl. Bact.* 25:87.
4. Duncan, D. W. & W. E. Razzell. 1967. Klebsiella biotypes among coliforms isolated from forest environments and farm produce. *Appl. Microbiol.* 24:933.
5. Brown, C. & R. J. Seidler. 1973. Potential pathogens in the environment: *Klebsiella pneumoniae*, a taxonomic and ecological enigma. *Appl. Microbiol.* 25:900.
6. Geldreich, E. E. 1966. Sanitary Significance of Fecal Coliforms in the Environment. U.S. Department of Interior, Fed. Water Pollut. Control Admin. Publ. WP-20-3.
7. Tennant, A. D., J. E. Reid & L. J. Rodewell. 1959. An Evaluation of the EC Confirmation Test for the Estimation of *Escherichia coli*. Densities in Seawater and Shellfish. Laboratory of Hygiene Manuscript Report No. 59-6.
8. Menon, A. S. 1974. Bacteriological Study of the Annapolis Basin. Environmental Protection Service Surveillance Report EPS-AR-73-17.
9. Dufour, A. P. & V. J. Cabelli. 1976. Characteristics of Klebsiella from textile finishing plant effluents. *J. Water Pollut. Control Fed.* 48:872.
10. Bordner, R. & J. Winter, eds. 1978. Microbiological Methods for Monitoring the Environment. Water and Wastes. EPA-600/8-78-017. U.S. EPA, Cincinnati, Ohio.

11. Bacteriological Analytical Manual. 1978. U.S. Food and Drug Administration, Washington, D.C.
12. Hunt, D. A. & J. Springer. 1978. Comparison of two rapid test procedures with the standard EC test for recovery of fecal coliform bacteria from shellfish-growing waters. *J. Ass. Off. Anal. Chem.* 61:1317.

3.7 BIBLIOGRAPHY

EIJKMAN, C. 1904. Die Gärungsprobe bei 46° als Hilfsmittel bei der Trinkwasseruntersuchung. *Zentr. Bakteriol. Parasitenk.* Abt. I, Orig. 37:742.

JORDAN, H. E. 1932. Brilliant green bile for coli-aerogenes group determinations. *J. Amer. Water Works Ass.* 24:1027.

HOSKINS, J. K. 1933. The most probable number of *B. coli* in water analysis. *J. Amer. Water Works Ass.* 25:867.

HOSKINS, J. K. 1934. Most probable numbers for evaluation of *coli-aerogenes* tests by fermentation tube method. *Public Health Rep.* 49:393 (Reprint 1621).

KELLY, C. B. 1940. Brilliant green lactose bile and the standard methods completed test in isolation of coliform organisms. *Amer. J. Pub. Health.* 30:1034.

DALLA VALLE, J. M. 1941. Notes on the "most probable number" index as used in bacteriology. *Amer. J. Pub. Health.* 56:229.

HAJNA, A. A. & C. A. PERRY. 1943. Comparative study of presumptive and confirmation media for bacteria of the coliform group and for fecal streptococci. *Amer. J. Pub. Health* 33:550.

MCCRADY, M. H. 1943. A practical study of lauryl sulfate tryptose broth for detection of the presence of coliform organisms in water. *Amer. J. Pub. Health.* 33:1199.

CLARK, H. F., E. E. GELDREICH, H. L. JETER & P. W. KABLER. 1951. The membrane filter in sanitary bacteriology. *Pub. Health Rep.* 66:951.

KABLER, P. W. 1954. Water examination by membrane filter and MPN procedures. *Amer. J. Pub. Health.* 44:379.

TAYLOR, E. W., et al. 1955. *J. Inst. Water Engrs.* 9:248.

GELDREICH, E. E., H. F. CLARK, C. B. HUFF & L. C. BEST. 1965. Fecal-coliform-organism medium for the membrane filter technic. *J. Amer. Water Works Ass.* 57:208.

BECK, W. J. 1968. Methodology for Testing for Fecal coliform organisms from the marine environment. *In* Proc. Symp. on Fecal Coliform Bacteria in Water and Wastewater. Bur. of Sanit. Eng., California State Dept. of Pub. Health, Berkeley.

CLARKE, N.F., 1975. Advantages and limitations of the membrane filter procedure. *Water & Sew. Works* 104:385.

ROSE, R. E., E. E. GELDREICH & W. LITSKY. 1975. Improved membrane filter method for fecal coliform analysis. *Appl. Microbiol.* 29:532.

LIN, S. D. 1977. Comparison of membranes for fecal coliform recovery in chlorinated effluents. *J. Wat. Pollut. Control Fed.* 49:225.

STUART, D. G., G. A. MCFETERS & J. E. SCHILLINGER. 1977. Membrane filter technique for the quantification of stressed fecal coliforms in the aquatic environment. *Appl. Environ. Microbiol.* 34:42.

TOBIN, R. S. & B. J. DUTKA. 1977. Comparison of the surface structure, metal binding and fecal coliform recoveries of nine membrane filters. *Appl. Environ. Microbiol.* 34:69.

CHAPTER 4

BIOASSAY PROCEDURES FOR SHELLFISH TOXINS

John E. Delaney

4.1 INTRODUCTION

A. Paralytic Shellfish Poison (PSP)

Several species of marine dinoflagellates synthesize neurotoxins that accumulate in mollusks during certain times of the year. Epidemiological studies indicate that these toxins can cause paralysis and death in humans. Although the taxonomy has not been completely settled, it appears that *Gonyaulax tamarensis var excavata* (also designated *G. excavata* and *G. tamarensis*) is the causative organism of PSP on the Atlantic coast. On the Pacific coast *G. tamarensis* and *G. catenella* have been reported; however, the principal species causing PSP is considered to be *G. catenella*. These dinoflagellates recently were transferred to the new genus *Protogonyaulax*.[1]

The occurrence of PSP in shellfish is difficult to predict. It has not been possible to correlate the occurrence of PSP in shellfish with any specific ecological or environmental condition. Careful monitoring of suspect areas is the only reliable method of control. Generally, on the Atlantic coast PSP seldom occurs in shellfish south of Cape Cod; however, from Cape Cod to Maine and in Canada to the north, occurrences are sporadic, with toxic shellfish found in some areas and not in others.

On the Pacific coast there have been frequent, often annual, occurrences of PSP in mussels in California. Toxic shellfish have been found infrequently along the Oregon coast, the southern shore of Washington, the outer straits of Juan de Fuca, and the shores of

British Columbia. The occurrence of PSP in shellfish is perennial in most Alaskan waters.

PSP is quantified by a bioassay using the mouse unit (MU) as the unit of measurement.[2] The MU was defined as the minimum amount of poison required to kill a mouse weighing 20 g in 15 min when 1.0 mL of shellfish extract is injected intraperitoneally (IP). The bioassay procedure has been modified by using purified saxitoxin as a reference standard and converting mouse units to micrograms (µg) of toxin.[3] This bioassay procedure and the expression of results in terms of toxin concentration is presented here, but it is recognized that bioassay results may represent total toxicity from several toxins present.

The Association of Official Agricultural Chemists first recognized the bioassay procedure for PSP as an official method in 1960.[4]

B. *Ptychodiscus brevis* Toxin(s)

The marine dinoflagellate *Ptychodiscus brevis* (Davis) Steidinger (= *Gymnodinium breve*) has been isolated and identified from coastal waters of the Gulf of Mexico and the Caribbean. It produces several neurotoxic and hemolytic substances.[5] Filter-feeding bivalve shellfish, such as oysters, clams, and coquinas, can accumulate the toxin(s) and become toxic to humans. Symptoms of shellfish poisoning caused by *P. brevis* include loss of coordination, gastrointestinal distress, numbness, and tingling sensations in mouth, hands, and feet. Although shellfish poisonings were reported during an 1880 red tide,[6] the causes were not generally recognized until more recently.[7] To date no human deaths have been attributed to this toxigenic agent, although several individuals have become seriously ill.

P. brevis toxin(s) can be detected in shellfish meats by a standardized mouse bioassay.[7,8] The bioassay is based on the relationship of dose to death time of mice injected IP with lipid-soluble crude toxin residues extracted from shellfish with diethyl ether. Because no purified standard is available, relative toxicity is expressed in MU. One MU is that amount of crude toxin that, on the average, will kill 50% of the test animals (20 g mice) in 930 min.

4.2 METHOD FOR PSP

A. Apparatus

Electric blender.
Torsion balance, 0.5 g sensitivity.
Hot plate.
Centrifuge.
pH meter.
Stop watch, mechanical or electronic, registering to at least 1 s.
Glassware: 800 mL beakers, 200 mL graduated cylinder, and 100 mL volumetric flask.
Plastic stirring rods.
Disposable syringes with 26-gauge needles.

B. Reagents

Hydrochloric acid, $0.18N$: dilute 15 mL conc HCl to 1 L with distilled water.
Hydrochloric acid, $5N$: dilute 41.7 mL conc HCl to 100 mL with distilled water.
Sodium hydroxide, $0.1N$: dissolve 4.6 g NaOH in 1 L distilled water.

C. Special Materials

PSP standard solution (100 µg/mL): This solution is available from HHS/PHS/FDA Special Projects Section, 1090 Tusculum Ave., Bldg. T-10, Cincinnati, Ohio 45226. It is acidified with HCl (to about pH 3), contains 20% ethyl alcohol as a preservative, and is stable indefinitely in a cool place. After opening the vial, store it in an airtight container under normal refrigeration temperatures to prevent evaporation and to maintain stability.

Mice: Use healthy mice, weighing 19 to 21 g obtained from a stock colony. Animals weighing 17 to 19 and 21 to 23 g may be used in the absence of animals in the most desirable weight range. Do not reuse surviving mice. Because mice often eat excessively and may store 1 to 2 g of food in the digestive tract, feed them sufficiently but do not overfeed.

D. Standardization of Bioassay

1. *Procedure*

Prepare PSP reference solution (1 µg saxitoxin/mL) by pipetting 1.00 mL standard solution into a 100 mL volumetric flask and make up to volume with distilled water acidified with HCl (to about pH 3). This solution is stable for several weeks if stored at 3 to 4°C and if the final pH is between 2.0 and 4.0.

Dilute 10 mL portions of 1 µg/mL reference solution with 10, 15, 20, 25, and 30 mL distilled water, respectively, until IP injection of 1 mL volumes into a few test mice cause a median death time of 5 to 7 min; maintain pH of dilutions at between 2.0 and 4.0. Figures 4:1, 4:2, and 4:3, respectively, show the proper techniques of positioning, holding, and injecting a mouse for the PSP bioassay.

Test additional dilutions in 1 mL increments of distilled water, e.g., if 10 mL reference solution diluted with 25 mL distilled water kills mice in 5 to 7 min, then test dilutions of 10 mL reference solution plus 24 mL distilled water and of 10 mL reference solution plus 26 mL distilled water. Maintain all dilutions between pH 2.0 and 4.0. Record injection and death times to the nearest 5 s interval; thus 7 s is rounded off to 5 s, and 8 s to 10 s. Death time is the elapsed time between completion of the injection and the last gasping breath of the mouse.

Weigh and record the weight of 10 mice to nearest 0.5 g and inject each with 1 mL of two, or preferably three, dilutions that cause median death times of 5 to 7 min. Record death times. If more than 3 mice in a group of 10 survive the injection of a particular dilution of the reference solution, repeat the injection with a new group of 10 mice. Before proceeding with the second group, investigate variables in the procedure that could have caused the initial results, such as leakage of the injected mixture from the mouse or failure to inject full volume of standard solution.

2. Calculations

Calculate median death time for each group of 10 mice used with each dilution. Discard results for any group of 10 mice giving a median death time <5 or >7 min. If any group of 10 mice gives a median death time between 5 and 7 min, include all groups of 10 mice used for that dilution in subsequent calculations even though some death times fall outside the desired range.

Using the death times for each mouse in each group in which the median death time falls between 5 and 7 min, determine the corresponding MU from Table 4:1. Using the weight of each mouse, determine the weight correction factor from Table 4:2. Multiply the MU by the weight correction factor to determine the values for corrected mouse units (CMU) per milliliter of the selected dilutions. Divide the calculated μg poison/mL in the selected dilutions by the associated CMU to obtain conversion factor (*CF*) values. Calculate the average of individual *CF* values and use the resultant value to check routine assays. This value represents micrograms of poison equivalent to one MU.

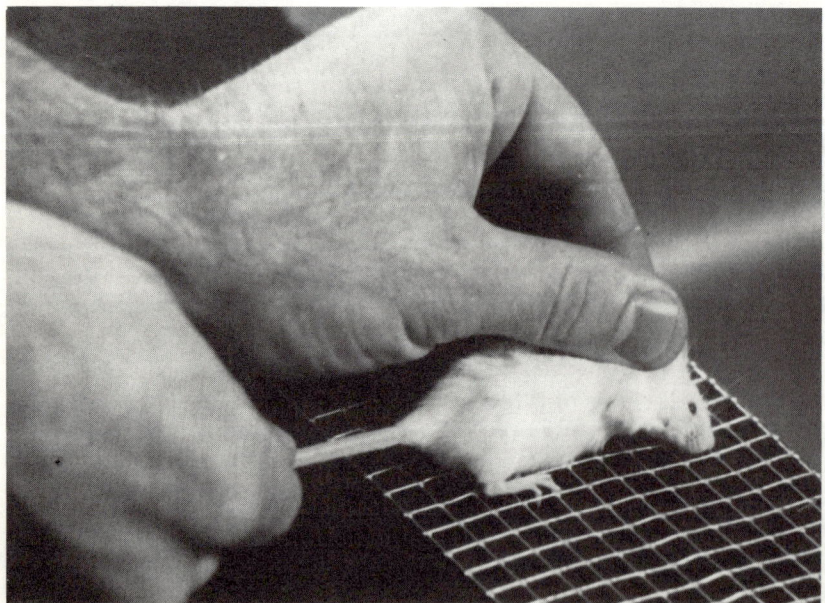

Figure 4:1—Positioning and grasping of mouse.

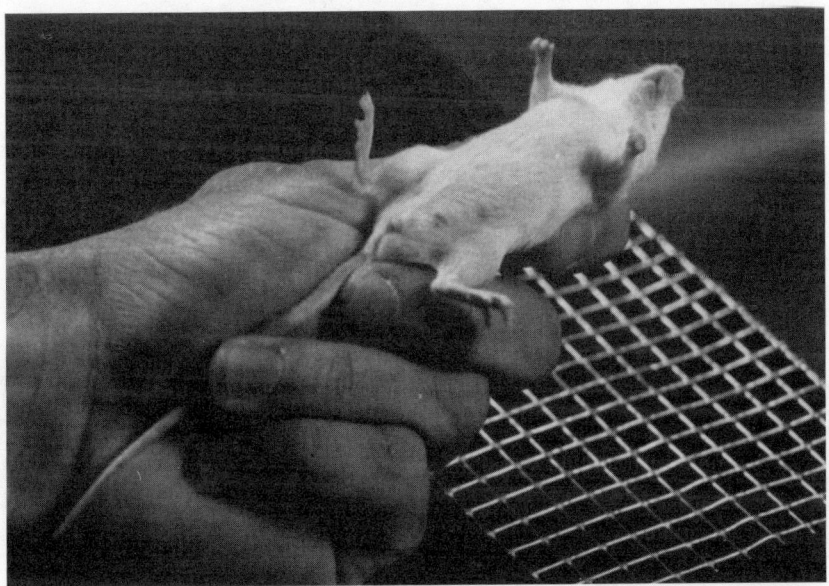

Figure 4:2—Pre-inoculation position.

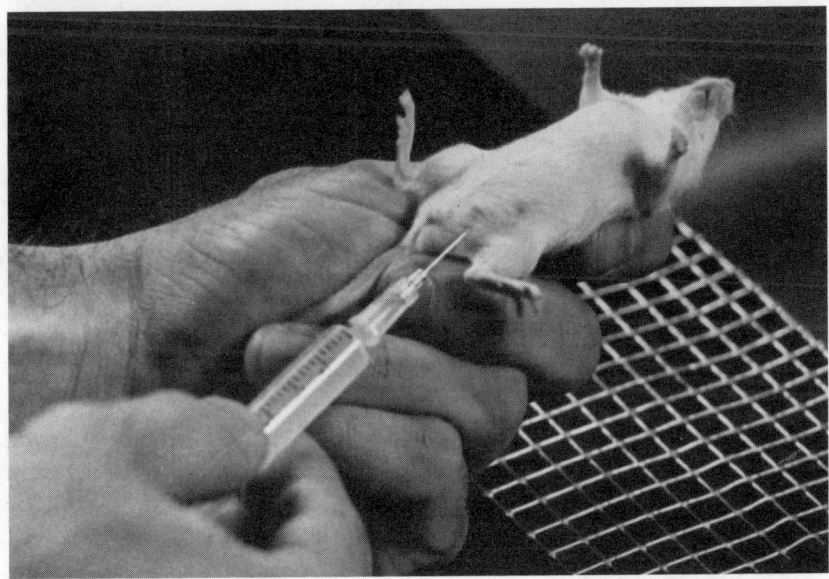

Figure 4:3—Correct needle angle for intraperitoneal (I.P.) injection.

Laboratory Procedures for Seawater and Shellfish

TABLE 4:1. DEATH TIME: MOUSE UNIT RELATIONS FOR PARALYTIC SHELLFISH POISON

Death Time*	Mouse Units	Death Time*	Mouse Units	Death Time*	Mouse Units	Death Time*	Mouse Units
1:00	100.0	3:00	3.70	55	1.96	30	1.13
10	66.2	05	3.57	5:00	1.92	10:00	1.11
15	38.3	10	3.43	05	1.89	30	1.09
20	26.4	15	3.31	10	1.86	11:09	1.075
25	20.7	20	3.19	15	1.83	30	1.06
30	16.5	25	3.08	20	1.80	12:00	1.05
35	13.9	30	2.98	30	1.74	13	1.03
40	11.9	35	2.88	40	1.69	14	1.015
45	10.4	40	2.79	45	1.67	15	1.000
50	9.33	45	2.71	50	1.64	16	0.99
55	8.42	50	2.63	6:00	1.60	17	0.98
2:00	7.67	55	2.56	15	1.54	18	0.972
05	7.04	4:00	2.50	30	1.48	19	0.965
10	6.52	05	2.44	45	1.43	20	0.96
15	6.06	10	2.38	7:00	1.39	21	0.954
20	5.66	15	2.32	15	1.35	22	0.948
25	5.32	20	2.26	30	1.31	23	0.942
30	5.00	25	2.21	45	1.28	24	0.937
35	4.73	30	2.16	8:00	1.25	25	0.934
40	4.48	35	2.12	15	1.22	30	0.917
45	4.26	40	2.08	30	1.20	40	0.898
50	4.06	45	2.04	45	1.18	60	0.875
55	3.88	50	2.00	9:00	1.16		

* In minutes and seconds.

Individual *CF* values may vary significantly within a single laboratory if techniques are not controlled rigidly. Regularly use the reference standard or a secondary standard, depending on the volume of assay work performed.

3. Use of standard with routine assay of shellfish

Check *CF* value periodically as follows: If shellfish are assayed less than once a week, determine *CF* value on each day assays are performed by injecting 5 mice with an appropriate dilution of working standard. If assays are made on several days during the week, check only once per week with a diluted standard such that the median

TABLE 4:2. CORRECTION TABLE FOR WEIGHT OF MICE

Mouse Weight, g	Mouse Units	Mouse Weight, g	Mouse Units
10	0.50	17	0.88
10.5	0.53	17.5	0.905
11	0.56	18	0.93
11.5	0.59	18.5	0.95
12	0.62	19	0.97
12.5	0.65	19.5	0.985
13	0.675	20	1.000
13.5	0.70	20.5	1.015
14	0.73	21	1.03
14.5	0.76	21.5	1.04
15	0.785	22	1.05
15.5	0.81	22.5	1.06
16	0.84	23	1.07
16.5	0.86		

death time falls between 5 and 7 min. The *CF* thus determined should agree within ±20% of the predetermined standard *CF*. If results do not agree, base the *CF* check on a 10-mouse group formed by injecting 5 additional mice with the same dilution of standard PSP solution and including the results with those from the original 5 mice. Inject a second group of 10 mice. The average *CF* value obtained from the two groups of 10 mice represents a new *CF* value.

Repeated checks of the *CF* value ordinarily produce consistent results within ±20%. If wider variations are found frequently, investigate the possibility of uncontrolled or unrecognized variables in the method before proceeding with routine assays.

E. Preparation for Analysis of Samples

1. *Sample collection*

Collect samples as directed in Section 2.2B. Unless fresh shellstock can be analyzed promptly, separate shellfish meats in the field as directed below and preserve 150 to 200 g drained meat in 100 mL 0.18N HCl. Protect acidified meat sample from leaking and refrigerate until analysis is started.

LABORATORY PROCEDURES FOR SEAWATER AND SHELLFISH

2. **Preparation of samples**

a. Clams, oysters, and mussels

Thoroughly clean outside of shellfish with fresh water. Cut adductor muscles and open. Rinse inside with fresh water to remove sand and other foreign material. Shuck meat from shell being careful not to cut or damage body of mollusk. Collect approximately 150 g meat on a No. 10 sieve without layering, and let drain for 5 min. Discard pieces of shell. Grind drained meat in a blender until homogeneous.

b. Scallops

Separate edible portion (adductor muscle) for analysis. Drain and grind as in a.

c. Canned shellfish

Transfer entire contents of can (meat and liquid) into blender and blend until homogeneous. For large cans, drain meat and collect all liquid. Determine weight of meat and volume of liquid. Recombine portions of each in proportionate quantities. Blend recombined portions until homogeneous.

d. Acid preserved shellfish meat

Drain acid solution from shellfish and save both fractions. Grind drained meat until homogeneous.

3. **Extraction of PSP**

Weigh 100 g homogenized meat into a tared 800 mL beaker. Add 100 mL 0.18N HCl (or the saved preserving acid from samples that had been preserved), stir thoroughly, and check pH with a wide-range indicator, phenol blue, or Congo red paper or a pH meter; pH should be between 2.0 and 4.0. If necessary, adjust pH by adding 5N HCl or 0.1N NaOH dropwise and stir to prevent local alkalinization and destruction of the poison.

Boil mixture gently for 5 min and let cool to room temperature. When cool, check pH and adjust to 2.0 to 4.0, if necessary. Transfer mixture to graduated cylinder and dilute to 200 mL with distilled water. Return mixture to beaker, stir to homogeneity, and let settle for several min. Transfer supernatant into centrifuge tube and centrifuge for 5 min at 3000 rpm. Decant clarified extract into clean sample bottle.

F. Mouse Bioassay Test

1. Procedure

Weigh and record weight of three mice for each sample to be analyzed. Inject each of the three test mice IP with 1 mL of centrifuged extract. This is the most critical step of the entire bioassay. If the injections are not made directly into the peritoneal cavity, the death time will not be reproducible. Reject any mouse from which more than one drop of injected mixture leaks. Activate a stopwatch at time of injection and observe mice carefully for time of death as indicated by the last gasping breath. Record death time for each mouse. If the median death time is less than 5 min, dilute extract with $0.01N$ HCl and inject another set(s) of mice to obtain death times between 5 and 7 min. If the median death time of undiluted extract is greater than 7 min, it may be used to determine sample toxicity.

2. Calculation of PSP concentration ($\mu g/100$ g shellfish meat)

Determine the MU/mL of extract that corresponds to observed death times from Table 4:1. Calculate the CMU by multiplying the MU corresponding to the death time for each mouse by the weight correction factor obtained from Table 4:2. Table 4:3 presents an example of the data generated during a PSP bioassay on an undiluted extract with a mouse strain having a CF value of 0.20 μg saxitoxin equivalent/MU.

TABLE 4:3. SAMPLE CALCULATIONS

Mouse Weight g	Dilution	Death time min:s	Mouse units MU	Weight corr. factor	Corrected mouse units CMU
18.5	1×	5:30	1.74	0.95	1.65
20.0	1×	6:00	1.60	1.0	1.60
17.0	1×	5:25	1.77	0.88	1.56

Use the median CMU/mL for the three bioassays to determine µg poison/100 g shellfish meat in accordance with the following equation:

µg PSP/100 g of shellfish meat
$$= \text{Median CMU/mL} \times CF \times DF \times 200$$

where CF and DF represent the predetermined conversion factor and dilution factor, respectively. The data presented in Table 4:3 yield the following:

$$\text{PSP} = 1.60 \times 0.20 \times 1 \times 200 = 64 \ \mu g/100 \ g.$$

Any value greater than 80 µg/100 g is considered hazardous and the shellfish are unsafe for human consumption.

4.3 METHOD FOR *Ptychodiscus brevis* TOXIN(S)

A. Apparatus

Electric blender.
Analytical balance.
Hot plate.
Explosion-proof centrifuge with 250 mL centrifuge cups and covered with foil.
Explosion-proof chemical hood.
1000 mL separatory funnel.
400 mL beakers.
Disposable syringes with 26-gauge needles (do not reuse).
Stopwatch, mechanical or electronic, registering to at least 1 s.

B. Reagents

Hydrochloric acid, HCl, conc.
Sodium chloride, NaCl.
Cottonseed oil.*
Diethyl ether, anhydrous, AR (ACS). Use ether only from a previously unopened container—peroxides reduce apparent toxicity.

---
* Household cottonseed oil used for cooking is acceptable.

C. Test Animals

Use healthy albino male mice of the Swiss-Webster strain weighing 20 ± 1 g. Mice weighing between 15 and 25 g may be used. Do not reuse surviving mice. Because mice often eat excessively and may store 1 to 2 g food in the digestive tract, feed them sufficiently but do not overfeed.

D. Preparation for Analysis of Samples

1. Sample collection.

Collect samples as directed in Sections 2.2B and 4.2E.

2. Preparation of samples.

Clean, shuck, and drain shellfish as described in Section 4.2E. The number of shellfish required for 100 g of homogenate varies from 2 large clams to 8 or 10 small oysters. Homogenize shellfish meats in an electric blender at high speed for 5 min. Weigh 100 g of the homogenate into pre-weighed 400 mL beaker and add 5 g NaCl and 1 mL conc HCl. Stir well. Heat mixture to boiling and cook for 5 min; stir frequently. Let cool to room temperature and transfer to a 1000 mL separatory funnel. Rinse beaker with ether; add rinse to separatory funnel.

Perform all subsequent steps under an explosion-proof hood.

Add 100 mL ether to homogenate, stopper, and shake vigorously (venting frequently) for 5 min. Centrifuge at 2000 rpm for 15 min. Carefully decant upper clear yellow ether phase into a 1000 mL separatory funnel, keeping solids in the centrifuge bottle. Repeat extraction three more times until the total amount of ether used is 400 mL.

Drain off and discard any bottom layer containing small shellfish pieces and/or water emulsion so that only the ether phase remains. Transfer ether extract to a 400 mL beaker pre-weighed to the nearest 0.01 g. Let ether evaporate in air under the hood until no trace of ether fumes is discernible. An oil residue, which is the crude toxin extract, will remain. Cover tightly and freeze for later bioassay.

LABORATORY PROCEDURES FOR SEAWATER AND SHELLFISH

E. Mouse Bioassay Test

1. Procedure

Bring weight of crude toxin residue to 9.17 g with cottonseed oil. The volume of oil and toxin mixture represents 10 mL. Thoroughly mix and break up remaining pieces of insoluble matter as small as possible with a stirring rod.

Slowly fill syringe (with needle in place) with 1 mL of residue–cottonseed oil mixture. Carefully inject 1 mL into each of two weighed mice IP on the ventral side anterior to the hind leg. If more than 1 drop of injected mixture leaks from the mouse, reject the mouse and inject another. Record time of injection. If the two mice survive 2 h, inject three more mice with 1 mL of residue–cottonseed oil mixture OR if the two mice die in less than 2 h, make dilutions until the injection solution causes the death of two mice in 2 to 6 h. (Note: the recommended dilution is 1:1.25, made by adding 2 mL cottonseed oil to the remaining 8 mL of residue-cottonseed oil mixture.) Repeat dilutions if necessary. When the correct dilution is found, inject three more mice.

Observe mice continuously for 6 h. The death time is the time elapsed from injection to the last breath of a mouse. The eyes will darken immediately upon death. If mice survive 6 h, hold them for a total of 24 h. For a 6-h continual observation period the lower limit of the assay's sensitivity is 20 MU/100 g shellfish. Extending the continuous observation period to 15.5 h will increase the assay's sensitivity to 10 MU/100 g shellfish. If mice die following continuous observation but within the 24 h period, toxin is present in low quantity.

During an assay, death may not occur but physiological signs of *P. brevis* toxicity may be observed. The more common signs of low toxicity (nonlethal) are weakness of limbs, imbalance, occasional respiratory spasms, and prolonged lethargy. Acute toxic signs include front and hindquarter paralysis resulting in instability, labored breathing, prostration, or hyperactivity. To observe nontoxic behavior for comparison, inject two mice with only cottonseed oil to serve as controls.

2. Calculations

Using Table 4:4, calculate the corrected MU using the following formula:

BIOASSAY PROCEDURES

MU × Weight Correction × Dilution = MU/100 g shellfish

If additional dilutions were not made, use the dilution factor of 10—based on the initial addition of cottonseed oil to 10 mL. If additional dilutions were made, multiply their factors also. Interpolate death times and weight corrections that fall between table values.

Example: If a 22.3 g mouse died in 390 min, using original dilution, then the MU = 1.9 × 1.14 × 10 = 21.7.

If mice die after continuous observation, calculate MU as though death occurred at the end of the continuous observation period and report results as an indeterminate value of "less than" (<), i.e., less

TABLE 4:4. RELATIONSHIP OF DOSE TO DEATH TIME AND WEIGHT OF MICE INJECTED WITH *PTYCHODISCUS BREVIS* TOXIN(S) EXTRACTED FROM SHELLFISH

Death time min (20 g mice)	Mouse units MU/mL	Mouse Weight Correction	
		Mouse Weight, g	Correction Factor
8	10.0	15	0.69
10	9.0	16	0.75
12	8.0	17	0.81
14	7.0	18	0.87
16	6.0	19	0.94
18	5.0	20	1.00
20	4.5	21	1.06
30	4.0	22	1.12
38	3.8	23	1.18
45	3.6	24	1.24
60	3.4	25	1.30
83	3.2	26	1.36
105	3.0		
140	2.8		
180	2.6		
234	2.4		
300	2.2		
360	2.0		
435	1.8		
540	1.6		
645	1.4		
780	1.2		
930	1.0		

than the sensitivity of the test for that time period. If mice survive the 24-h period, assign a value of <10 MU because the assay's lowest, reproducible sensitivity is 10 MU/100 g shellfish meat.

Calculate the mean MU if 100% mortality occurs and death times are determinate OR determine the median MU if less than 100% mortality is observed *or* if death times are indeterminate. When reporting indeterminate toxicity, note the number of mice that died in 24 h OR if no mice die in 24 h, report toxin undetectable (<10 MU/100 g shellfish meat).

Consider any detectable level of toxin per 100 g shellfish meats as rendering the shellfish potentially unsafe for human consumption.

4.4 REFERENCES

1. Taylor, F. J. R. 1979. The toxigenic gonyaulacoid dinoflagellates. *In* Taylor, D. L. & H. H. Seliger, eds., Toxic Dinoflagellate Bloom. Elsevier-North Holland, Inc., New York.
2. Sommer, H. & K. F. Meyer. 1937. Paralytic shellfish poisoning. *Arch. Path.* 24:560.
3. Schantz, E. J., E. F. MacFarren, M. L. Schaeffer & K. H. Lewis. 1958. Purified shellfish poison for bioassay standardization. *J. Ass. Off. Anal. Chem.* 41:160.
4. Association of Official Agricultural Chemists. 1965. Paralytic Shellfish Poison, Biological Method (18). Official Methods of Analysis of the Association of Official Agricultural Chemists, 10th ed., Washington, D.C., pp. 282–284.
5. Mende, T. E. & D. G. Baden. 1978. Red tides—ecological headaches and research tools. *Trends in Biological Sciences* 3:209.
6. Walker, S. T. 1884. Fish mortality in the Gulf of Mexico. *Proc. U.S. Nat. Mus.* 6:105.
7. McFarren, E. F., H. Tanabe, F. J. Silva, W. E. Wilson, J. E. Campbell & K. H. Lewis. 1965. The occurrence of ciguatera-like poison in oysters, clams and *Gymnodinium breve* cultures. *Toxicon* 3:111.
8. Cummins, J. M. & W. F. Hill, Jr. 1969. Special Report: Method for the Bioassay of *Gymnodinium breve* Toxin(s) in Shellfish. Gulf Coast Marine Health Sciences Laboratory, Dauphin Island, Alabama 36528.

4.5 BIBLIOGRAPHY

MEDCOF, J. C., A. H. LEIM, A. B. NEEDLER, A. W. H. NEEDLER, J. GIBRARD & J. NAUBERT. 1947. Paralytic shellfish poisoning on the Canadian Atlantic Coast. *Bull. Fish. Res. Can.* 75:1.

NEEDLER, A. B. 1949. Paralytic shellfish poisoning and *Gonyaulax tamarensis*. *J. Fish. Res. Board Can.* 7:490.

RIEGEL, B., D. W. STANGER, D. W. WIKHOLM, J. D. MOLD & H. SOMMER, 1949. Paralytic shellfish poison. V. The primary source of the poison, the marine plankton organism *Gonyaulax catenella*. *J. Biol. Chem.* 177:7.

STEPHENSON, N. R., H. I. EDWARDS, B. F. MCDONALD & L. I. PUGSLEY. 1955. Biological assay of the toxin from shellfish. *Can. J. Biochem. Physiol.* 33:849.

MCFARREN, E. F., 1959. Report on collaborative studies of the bioassay for paralytic shellfish poison. *J. Ass. Off. Agric. Chem.* 42:263.

BURKE, J. M., J. MARCHISOTTO, J. J. A. MCLAUGHLIN & L. PROVASOLI. 1960. Analysis of the toxin produced by *Gonyaulax catenella* in axenic culture. *Ann. N. Y. Acad. Sci.* 90:837.

PRAKASH, A. 1963. Source of paralytic shellfish toxin in the Bay of Fundy. *J. Fish. Res. Board Can.* 20:983.

RAY, S. M. & D. V. ALDRICH. 1965. *Gymnodinium breve;* induction of shellfish poisoning in chicks. *Science* 148:1748.

PRAKASH, A. & F. J. R. TAYLOR. 1966. Red water bloom of *Gonyaulax acatenella* in the Strait of Georgia and its relation to paralytic shellfish toxicity. *J. Fish. Res. Board Can.* 23:1265.

SCHANTZ, E. J., J. M. LYNCH, G. VAYVADA, K. MATSUMOTO & H. RAPOPORT. 1966. The purification and characterization of the poison produced by *Gonyaulax catenella* in axenic cultures. *Biochemistry* 5:1191.

CUMMINS, J. M., A. A. STEVENS, B. E. HUNTLEY, W. F. HILL & J. E. HIGGINS. 1968. Some properties of *Gymnodinium breve* toxin(s) determined bioanalytically in mice. *In* Freudenthal, H. D., ed., *Drugs from the Sea*. Marine Technology Society, Washington, D.C.

STEVENS, A. A., J. M. CUMMINS & W. F. HILL, JR. 1968. A Method for Separation of Two Toxic Components from Cultures of *Gymnodinium breve*. *Tech. Report GCMHSL 68–3*, Gulf Coast Marine Health Sciences Laboratory, Dauphin Island, Alabama 36528.

CUMMINS, J. M. & A. C. JONES. 1969. Uptake and elimination of

Gymnodinium breve toxin(s) by the oyster *Crassostrea virginica*. *1968 Proceedings of the National Shellfish Association*, Vol. 59 (Abstract).

LOEBLICH, L. A. & A. R. LOEBLICH III. 1975. The organisms causing New England red tides: *Gonyaulax excavata*. *In* LoCicero, V. R., ed., Proceedings of the 1st International Conference on Toxic Dinoflagellate Blooms. Massachusetts Science and Technology Foundation, Wakefield, Massachusetts.

SCHANTZ, E. J., V. E. GHAZAROSSIAN, H. K. SCHNOES, F. M. STRONG, J. P. SPRINGER, J. O. PEZZANITE & J. CLARDY. 1975. Structure of saxitoxin. *J. Amer. Chem. Soc.* 97:1238.

SHIMIZU Y., M. ALAM, Y. OSHIMA & W. E. FALLON. 1975. Presence of four toxins in red tide infested clams and cultured *Gonyaulax tamarensis* cells. *Biochem. Biophys. Res. Comm.* 66:731.

ALAMA, M. I., C. P. HSU & Y. SHIMIZU. 1979. Comparison of toxins from three isolates of *Gonyaulax tamarensis* (Dinophyceae). *J. Physiol.* 15:106.

YENTSCH, C. M. & J. W. HURST. 1979. Reports of Investigations toward an Environmental Productive Index for Toxic Dinoflagellate Blooms. Bigelow Laboratory for Ocean Sciences and State of Maine Department of Marine Resources, West Boothbay Harbor, Maine.

CHAPTER 5

PROCEDURES FOR THE VIROLOGICAL EXAMINATION OF SEAWATER, SHELLFISH, AND SEDIMENT

Mark D. Sobsey

5.1 INTRODUCTION

Human enteric viruses, a large and heterogenous group, may be transmitted by the fecal-oral route, may be present in shellfish raising waters, and may accumulate in shellfish or sediments. Included are such viruses as polioviruses, coxsackieviruses, echoviruses and other enteroviruses, adenoviruses, reoviruses, rotaviruses, hepatitis A (infectious hepatitis) virus, and Norwalk-type gasteroenteritis viruses. There are more than 100 serologically distinct viruses in the group.

Enteric virus contamination of mollusks has been documented by recovery of viruses from shellfish[1-4] and by occurrence of shellfish-borne enteric virus disease outbreaks, such as hepatitis type A and viral gastroenteritis.[2,5-8] Recent epidemiological and microbiological findings have raised concerns about the reliability and validity of present coliform standards for shellfish and harvesting waters. Outbreaks of hepatitis type A and viral gastroenteritis may have been caused by shellfish obtained from waters that were approved for harvesting.[8,9] There is no consistent relationship between levels of enteric viruses and coliform bacteria in shellfish or the waters and sediments from which they were harvested.[1,3,4,10,11]

Growing public health concern about the virological quality of shellfish and the adequacy of coliform standards for shellfish and harvesting waters dictates the need for further studies on the occurrence, persistence, and fate of enteric viruses in shellfish and their habitat, and the relationships, if any, between enteric viruses and indicator bacteria. In this chapter, the most promising virological methods are described. It must be emphasized, however, that the use of these virus detection methods should be limited primarily to special circumstances such as the development of new approaches for assessing, controlling, or improving shellfish sanitary quality; investigation of shellfish-borne disease outbreaks; and other research-oriented studies. Routine virus monitoring of shellfish or their habitat is not recommended due to the technical complexity, lengthiness, high cost, and limitations of the detection and recovery methods.

The efficiency of methods included here may vary widely, depending on such factors as: variability in types, amounts, and conditions of viruses in the samples; characteristics, quality, and size of samples; and characteristics and conditions of virus recovery and assay procedures. It is probable that only a small and variable proportion of the total viruses present in field samples are being detected with current procedures. The methods have not been evaluated systematically through extensive collaborative and quality assurance studies.

5.2 GENERAL METHODS AND PROCEDURES

A. Sterilizing Apparatus, Materials, and Reagents

Sterilize all items that directly contact samples by autoclaving or by alternate procedures. Sterilize non-autoclavable materials by cold sterilization with ethylene oxide or by treating with 10 mg/L free chlorine solution at pH 7.0 for 30 min, and then rinsing or flushing with a sterile solution of 50 mg $Na_2S_2O_3$/L. Use aseptic techniques in all virus recovery operations to prevent extraneous microbial contamination.

B. Hydrogen Ion Concentration (pH)

See Chapter 2.3D. When making pH measurements on different samples, decontaminate the electrodes between measurements by immersing in $1N$ HCl for 5 min, rinsing with sterile distilled water, and recalibrating with buffer standards.

C. Sample Collection

See Chapter 2.2. For water samples containing residual chlorine, immediately add $Na_2S_2O_3$ to a final concentration of 50 mg/L.

D. Sample Storage

Process samples as soon as possible; hold samples for no more than 2 h at up to 25°C or for no more than 48 h at 2 to 10°C. Do not freeze unless samples are not to be processed within 48 h; freeze at $\leq -70°C$. Use the same storage conditions for processed samples (final concentrates) after treating with antibiotics. For long term storage, freeze and hold at $\leq -70°C$.

E. Decontamination of Processed (Concentrate) Samples

Control bacteria and mold contamination of concentrated samples by adding 1 part each of kanamycin-gentamycin or pencillin-streptomycin solution and nystatin B suspension or amphotericin B solution to 50 parts of sample (w/v). Incubate at room temperature for 2 h or at 37°C for 1 h, and refrigerate or freeze ($\leq -70°C$) until virus assay is made. Freezing aids in destroying bacterial contaminants. To check effectiveness of treatment, plate a small subsample on a general purpose medium such as plate count agar by the spread plate technique and incubate at 37°C for 24 to 48 h.

If contamination persists add 1/10 sample volume of chloroform, $CHCl_3$, and mix vigorously for 30 min at room temperature or homogenize for 1 to 2 min at 4 to 10°C. To separate phases let stand in refrigerator overnight or centrifuge at $\geq 1000 \times g$. Remove aqueous upper layer (sample) with a pipet. Remove dissolved $CHCl_3$ from sample by bubbling with filter-sterilized air for 15 min; remove residual traces of $CHCl_3$ by placing sample in sterile, shallow container, uncovered, in a lamimar flow hood or biological safety cabinet for several hours. To decontaminate shellfish samples substitute 1/2 the quantity of trichlorotrifluoroethane and treat as with $CHCl_3$.

5.3 ENTERIC VIRUS CONCENTRATION FROM SHELLFISH

A. Introduction and General Description

Three procedures are described here: Adsorption-elution-precipitation; elution-precipitation; and filtration-hydroextraction. These methods are capable of separating or extracting viruses from shellfish meats; producing a concentrated sample that is relatively non-toxic to cell cultures, and recovering 50% or more of the viruses initially present in the shellfish. A preferred method cannot be specified, therefore, make a preliminary evaluation of candidate methods for virus recovery from samples seeded with representative enteric viruses. In addition, consider the availability and cost of facilities, equipment, and materials and the ease, convenience, and length of processing time.

Three adsorption-elution-precipitation procedures are described (Figure 5:1).[1,12-14] In these methods, viruses initially are adsorbed to homogenized shellfish tissue, and possibly other precipitated solids, at low salt concentration (conductivity) and pH. In all three modifications (Schemes A, B, and C, Figure 5:1) the suspending medium is distilled water and the homogenate is adjusted to pH 4.5 to 5.0 with HCl. The homogenate is centrifuged at low speed and the supernatant is discarded. Viruses are eluted from the sediment by resuspending in various media: A—glycine-saline, pH 7.5 and about 8000 mg NaCl/L; B—glycine-saline, pH 7.5 with Na_3PO_4; and C—glycine-saline with polyelectrolyte, pH 9.5. After resuspending the solids, centrifuge samples and discard sediment. Precipitate viruses in the supernatant with acid at pH 4.0 to 4.5, either directly (Scheme B), after filtration (Scheme A), or after adding beef extract to a final concentration of 1.5% (Scheme C). The virus-containing precipitate is recovered by centrifuging and is dissolved in either phosphate buffer and adjusted to pH 7.2 to 7.4 (Schemes A and C) or distilled water and adjusted to pH 9.0 followed by pH 7.5 (Scheme B). In Scheme A, the sample is treated with polyelectrolyte whereas in Schemes A and B it is centrifuged to remove additional solids and the supernatant is recovered. Finally, samples are treated with antibiotics and assayed for viruses.

Two modifications of elution-precipitation methods are included (Figure 5:2).[15,16] In both methods, the viruses are first eluted from shellfish meats by homogenizing (and sonicating-Scheme B) in al-

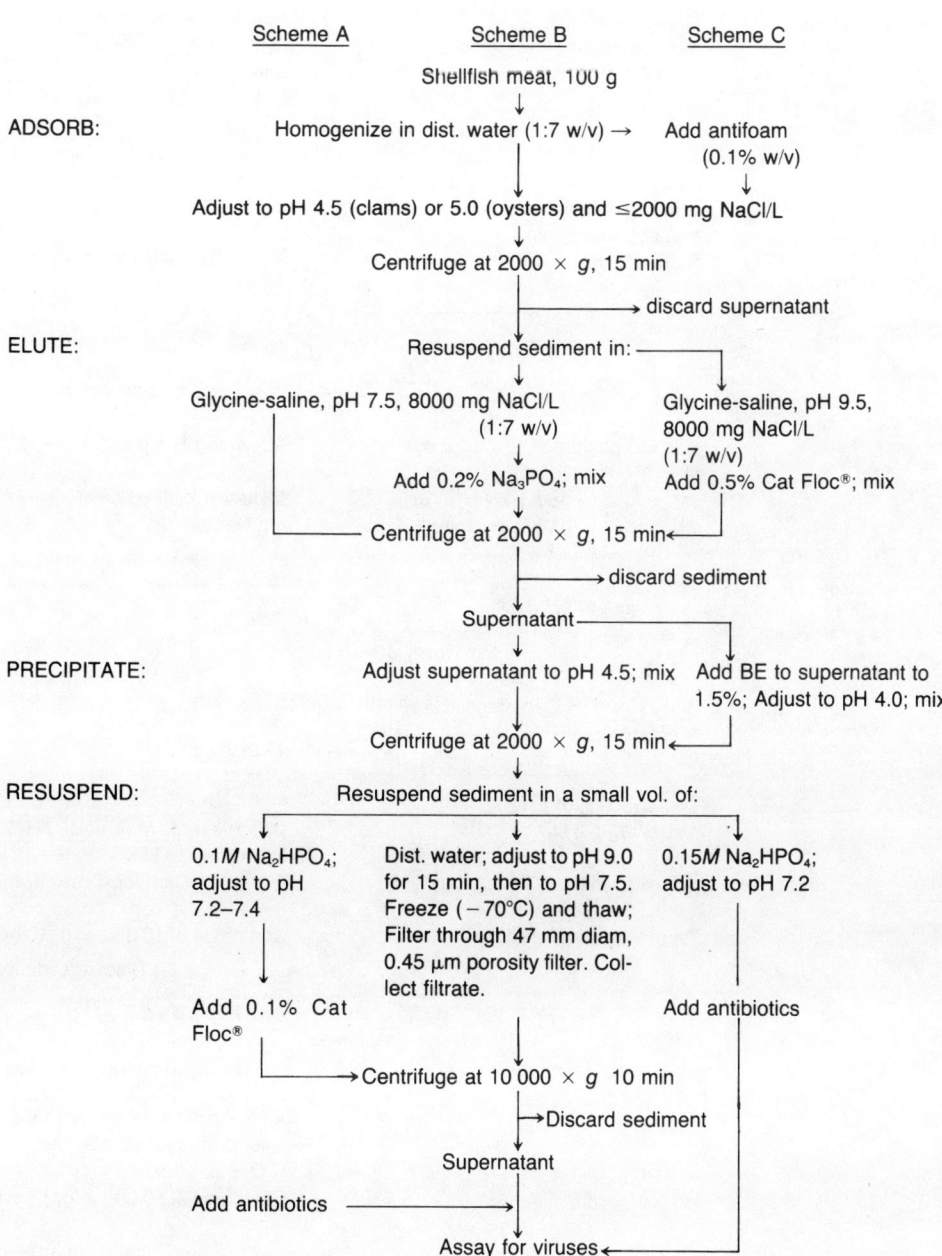

Figure 5:1—Virus recovery from shellfish by adsorption-elution-precipitation.

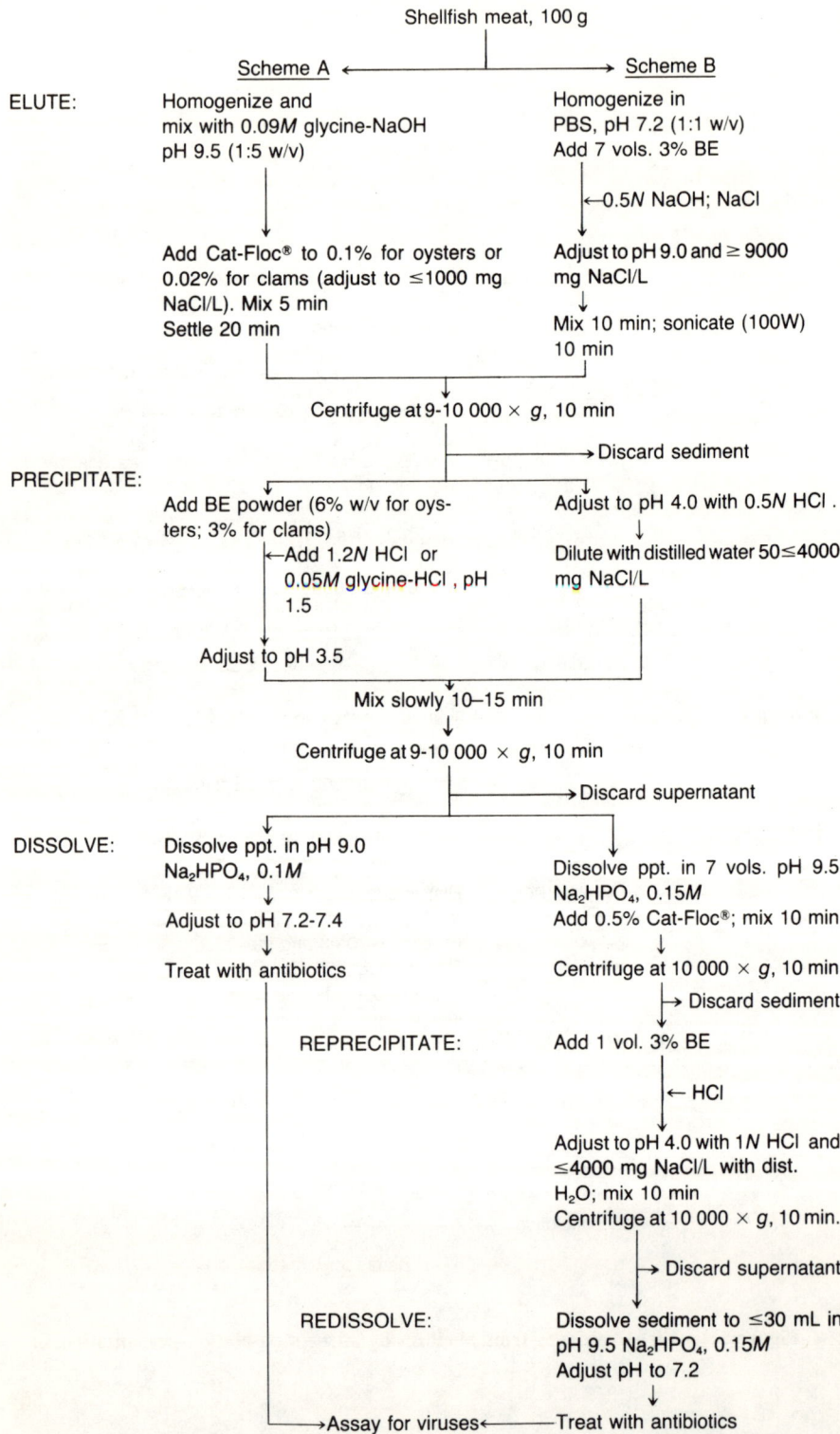

Figure 5:2—Elution-precipitation methods for virus recovery from shellfish.

kaline buffer solutions: glycine-polyelectrolyte (Scheme A) or phosphate-buffered beef extract (Scheme B). Remove shellfish solids by low-speed centrifuging. Precipitate viruses in the resulting supernatants by adding acid. Recover the precipitate by low-speed centrifugation and dissolve in either alkaline phosphate buffer and neutralize for virus assay (Scheme A) or dissolve in alkaline phosphate buffer with polyelectrolyte (Scheme B). In Scheme B centrifuge the sample and concentrate viruses further by adding beef extract, reducing conductivity, and acid precipitating at pH 4.0. Collect the precipitate by centrifugation, dissolve in alkaline phosphate buffer, neutralize, and assay for viruses.

The method for recovering viruses from shellfish by extraction from shellfish meats, clarification through a glass wool filter, and concentration by hydroextraction-dialysis against polyethylene glycol (PEG) is shown in Figure 5:3.[17] Viruses retained in the dialysis bag are recovered by washing the bag interior with alkaline buffer solution. The collected fluid is treated with diatomaceous earth to remove particulate and colloidal impurities. After removing the diatomaceous earth by centrifuging, the resulting supernatant is diluted in buffer and assayed for viruses.

B. Equipment and Materials

Scrubbing brushes, shucking knives and disinfectant solution. Disinfect with 70% ethanol, 10 mg sodium hypochlorite/L or $1N$ HCl.
Tared beaker or dish.
Blender. Sterilize container by autoclaving.
Homogenizer*. Sterilize container by autoclaving.
Balance.
Graduated cylinders: 100, 500, 1000, and 2000 mL.
H meter with combination electrode.
Beakers: 50, 100, 500, 1000, and 2000 mL.
Pipets: 1, 5, and 10 mL.
Centrifuge with rotors, buckets, bottles, and tubes for 25 to 500 mL samples. Capable of operating at $10\ 000 \times g$.

---

* Omnimixer, Dupont-Sorvall, or equivalent.

Laboratory Procedures for Seawater and Shellfish

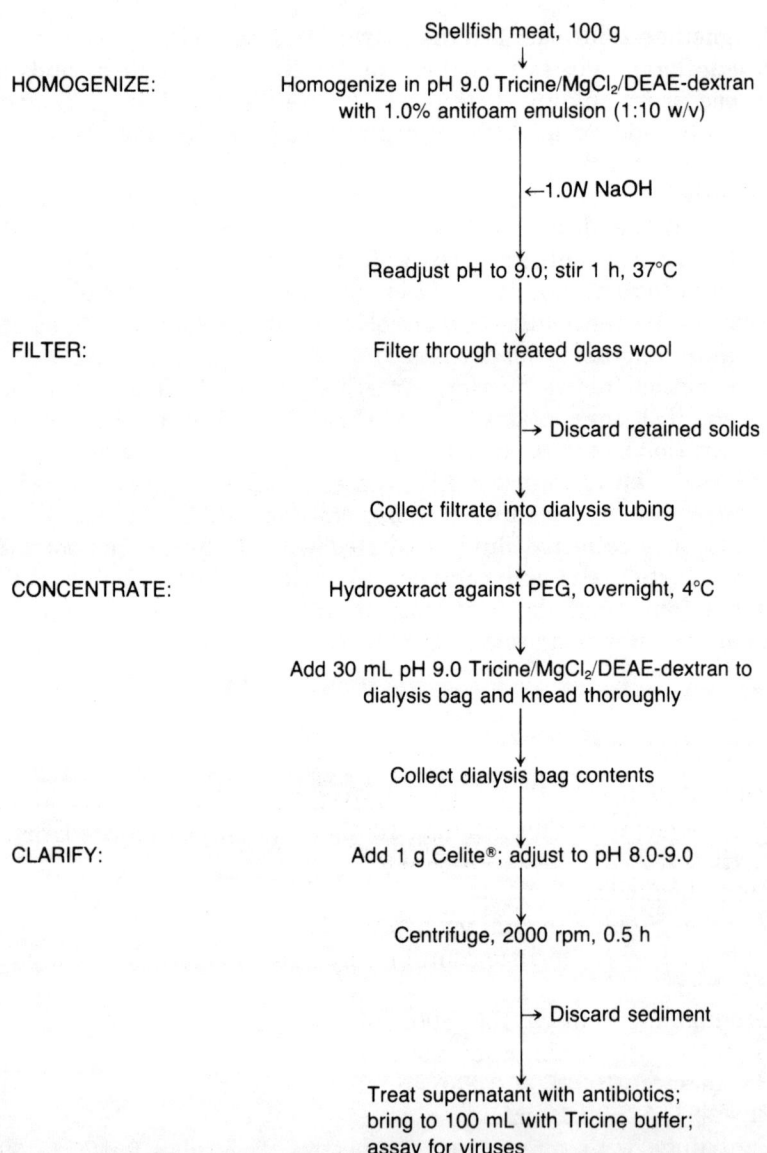

Figure 5:3—**Virus recovery from shellfish by glass wool-hydroextraction.**

Magnetic stirrer with stirring bars.

Vertical laminar-flow hood (Class II). Conduct all virus recovery and assay procedures under hood.

Conductivity meter with probe; preferably with meter calibrated in mg NaCl/L.

Sonicator: Probe type.

Reservoir or pressure vessel: Capacity 1 L, to hold samples for positive pressure filtration, equipped with inlet and outlet openings and pressure gauge.

Positive pressure source up to 276 kPa, with regulator: laboratory air line, positive pressure pump, or cylinder of compressed air or nitrogen.

Glass wool filter: layer 5 g glass wool, free of detergents or oils, into a 150-mm diam funnel to form a filter bed. Autoclave.

Dialysis tubing: seamless, regenerated cellulose, 4.8 nm average pore diam and 2.5 to 4.6 cm diam. Autoclave in 2 to 3 m lengths in distilled water.

Dialysis tubing clamps.

Sterile beaker or pan for dialysis tubing.

Wooden tongue depressor: Autoclave.

Rubber spatula or roller.*

C. Reagents

Hydrochloric acid, HCl, 10, 1.0, 0.5, and $0.1N$.

Sodium hydroxide, NaOH, 10, 1.0, 0.5, and $0.1N$.

Distilled water.

Antifoam emulsion.†

Phosphate buffered saline (PBS), pH 7.2 containing 8.0 g NaCl, 0.2 g KCl, 0.12 g KH_2PO_4, and 0.91 g Na_2HPO_4/L distilled water. Adjust to pH 7.2 with NaOH or HCl and sterilize by autoclaving.

Sodium chloride, NaCl, 16% (w/v).

$0.05M$ glycine-$0.14M$ saline (NaCl): Dissolve 3.75 g glycine and 8.18 g NaCl/L distilled water, adjust to pH 7.5 or 9.5 with NaOH, and autoclave.

---

* No. 14-245-21, Fisher Scientific Co., or equivalent.

† Antifoam Emulsion B, Dow Corning Corp., Midland, Mich., or equivalent.

Laboratory Procedures for Seawater and Shellfish

Beef extract*. To make 1.5, 3.0%, or 6% solutions, dissolve 15, 30, or 60 g powder/L distilled water, adjust pH to desired value with $1N$ HCl or $1N$ NaOH, and autoclave.

Disodium phosphate, Na_2HPO_4, 0.1 or $0.15M$: Dissolve 26.8 g (for $0.1M$) or 40.2 g (for $0.15M$) $Na_2HPO_4 \cdot 7H_2O$/L water, adjust pH to 9.0 to 9.5 with $1N$ HCl or $1N$ NaOH, and autoclave.

Trisodium phosphate, Na_3PO_4, 10%: Dissolve 100 g Na_3PO_4/L distilled water and autoclave. Use 1 part per 50 parts sample for an approximate 0.2% final concentration.

Polyelectrolyte†: Cationic, soluble polymer; polydimethyl-diallyl ammonium chloride. Dilute to desired concentration in distilled water and filter to sterilize on day of use.

$0.09M$ glycine-NaOH, pH 8.8: Dissolve 6.8 g glycine/L distilled water, adjust to pH 8.8 with $1N$ NaOH, and autoclave.

Antibiotics: Make stock solution of either kanamycin sulfate, 10 mg/mL and gentamycin, 2.5 mg/mL or penicillin G, 25 000 units/mL and streptomycin sulfate, 25 mg/mL. Make stock suspension of nystatin, 125 µg/mL or amphoteracin B, 50 µg/mL.

Polyethylene glycol-water (PEG-H_2O): Mix 7 parts of PEG (20 000 molecular weight, dry flakes) with 1 part sterile distilled water.

Tricine/$MgCl_2$/DEAE-dextran, pH 9.0 (TMD): Prepare $0.01M$ Tricine-$0.15M$ $MgCl_2$ solution by dissolving 1.79 g Tricine and 5.08 g $MgCl_2 \cdot 6H_2O$ in 1 L distilled water. Adjust to pH 9.0 with $1.0N$ HCl or $1.0N$ NaOH and autoclave. Dissolve 1 g DEAE-dextran in 100 mL distilled water and sterilize by filtration. To make TMD add 5 mL DEAE-dextran solution to 995 mL Tricine-$MgCl_2$ solution.

Tricine buffer, $0.01M$, pH 9.0: Dissolve 1.79 g Tricine in 1 L distilled water, adjust to pH 9.0 with $1N$ NaOH or $1N$ HCl, and autoclave.

Diatomaceous earth.** Autoclave.

Trichlorotrifluoroethane.††

---

* Lablemco Power, Oxoid USA, Inc., Columbia, Md., or equivalent.
† Cat-Floc® Calgon Corp., Pittsburgh, Penn.
** Celite®
†† Freon®

D. Procedures

1. *Adsorption-elution-precipitation procedures for shellfish*
  a. Adsorption

  Aseptically scrub and shuck enough shellfish to obtain 100 g or more meat and place in blender container. Add 7 parts cold distilled water (w/v) (Schemes A, B, and C) and blend to homogenize completely, about 2 min in a conventional blender. Transfer homogenate to a beaker with a magnetic stirring bar. Add antifoam emulsion to a final concentration of 0.1% (w/v) (Scheme C only). While mixing, adjust to pH 4.5 for clams and mussels (Schemes A, B, and C) or pH 5.0 for oysters (Schemes A, B, and C) with 1N HCl. Measure conductivity and, if necessary, adjust to the equivalent of ≤2000 mg NaCl/L by adding sterile distilled water. Centrifuge at 2000 × g (Schemes A, B, and C) for 15 min and discard supernatant.

  b. Elution

  Resuspend sediment by homogenizing for 0.5 to 1.0 min in 7 volumes glycine-saline and adjust to pH 7.5 (Schemes A and B) or pH 9.5 (Scheme C). While mixing, add 0.1% (v/v) Na_3PO_4 and readjust to pH 7.5 (Scheme B) or add 0.5% polyelectrolyte and mix 10 min (Scheme C). Centrifuge at 2000 × g for 15 min and collect supernatant.

  c. Acid precipitation

  Add 6.0% beef extract to a final concentration of 1.5% (w/v) (Scheme C only). While mixing, add 1.0N HCl and adjust to pH 4.5 (Schemes A and B) or pH 4.0 (Scheme C) and mix slowly for 10 min. Centrifuge at 2000 × g (Schemes A, B, and C) for 10 min and discard resulting supernatant. Dissolve sediment in enough 0.1M (Scheme A) or 0.15M (Scheme C) Na_2HPO_4, pH 9.0 to 9.5, to adjust to pH 7.2 or 7.4. For Scheme A, add polyelectrolyte to give a final concentration of 0.1% (1 volume 10% solution/100 volumes of sample) and let stand for 10 min. For Scheme B, dissolve sediment in sterile distilled water using 1/10th the sample volume before centrifuging, adjust to pH 9.0 with 1.0 and/or

0.1N NaOH, and let stand 15 min. For Schemes A and B centrifuge at 2000 × g for 15 min and collect supernatant. Adjust the Scheme B sample to pH 7.5 with 1.0 and/or 0.1N HCl, freeze at $-70°C$, thaw, and centrifuge at 1000 × g for 10 min. Filter through a 47-mm diam, 0.45 μm pore diameter membrane filter and collect filtrate.

For all Schemes, treat samples with antibiotics (see Section 5.2.E), freeze samples, and store at $-70°C$ until virus assay.

2. *Elution-precipitation procedures for shellfish*
a. Elution

Aseptically scrub and shuck enough shellfish to obtain 100 g or more meat and place in blender or homogenizer container (Scheme B). For Scheme B add a volume in mL cold PBS, pH 7.2, equal to the weight in g of shellfish. Blend to homogenize completely, about 2 min.

For Scheme A only, transfer homogenate to a beaker with a magnetic stirring bar and while mixing, add 5 parts (w/v) cold glycine-NaOH, pH 9.5, and 1% polyelectrolyte solution, to make final concentration of 0.1 and 0.02% (w/v) of the diluted homogenate, respectively, for oysters and clams. Readjust to pH 9.5 with 1.2N NaOH or HCl. Check conductivity and, if necessary, adjust to the equivalent of ≤1000 mg NaCl/L with sterile distilled water. Mix for 5 min and let stand for 20 min.

For Scheme B only, transfer homogenate to a beaker with a magnetic stirring bar and while mixing add 7 volumes (700 mL/100 g shellfish meat) cold 3% beef extract, pH 9.0. Adjust to pH 9.0 with 0.5N NaOH and measure conductivity. If necessary, add sufficient 16% NaCl to raise conductivity to ≥9000 mg NaCl/L and homogenize or sonicate (100 watts) for 10 min in an ice bath.

For both Schemes, centrifuge at 9 to 10 000 × g for 10 min and collect supernatant in a beaker.

b. Precipitation

For Scheme A only, add beef extract powder to a final concentration (w/v) of 6% for oysters or 3% for clams and mix to dissolve. For both Schemes, mix and add 0.5N HCl to adjust pH to 4.0 (Scheme B) or pH 3.5 (Scheme A). For Scheme B only, add sufficient sterile distilled water to adjust conductivity to the equivalent of ≤4000 mg NaCl/L. For both schemes, mix slowly for 10 to 15 min. Dissolve sediment either in enough $0.1M$ Na_2HPO_4, pH 9.0 to bring the pH 7.2 to 7.4 (Scheme A) or in 7 volumes $0.15M$ Na_2HPO_4, pH 9.5 (Scheme B).

c. Reprecipitation (Scheme B)

For Scheme B add 0.5% (v/v) polyelectrolyte. Mix 10 min, centrifuge at 9 to 10 000 × g for 10 min, and collect supernatant. While mixing, add an equal volume of 3% beef extract and adjust to pH 4.0 with $0.5N$ HCl. Add sufficient sterile distilled water to adjust conductivity to the equivalent of ≤4000 mg NaCl/L. Mix slowly for 10 min and centrifuge at 10 000 × g. Discard supernatant. Redissolve precipitate in a small volume (≤30 mL for an initial 100 g sample) of $0.15M$ Na_2HPO_4 and, if necessary, adjust pH to 7.2 with $0.5N$ HCl.

d. Antibiotic treatment

Treat final concentrates from either scheme with antibiotics (Section 5.2E) and store (Section 5.2D) until virus assay.

3. *Glass wool filtration-hydroextraction procedure for shellfish*
a. Shellfish homogenization

Aseptically scrub and shuck enough shellfish to obtain 100 g or more meat and place in blender container. Add 9 parts (w/v) cold Tricine/$MgCl_2$/DEAE-dextran solution and 1 part antifoam emulsion/100 parts diluted sample. Blend to completely homogenize, about 15 to 20 s in a conventional blender. Check homogenate pH after blending and, if necessary, readjust to pH 9.0 with $1N$ NaOH. Stir homogenate for 1 h at 37°C to emulsify lipids and free viruses from solids.

b. Glass wool filtration

Pour sample through a glass wool filter funnel pretreated with 100 mL Tricine, pH 9.0. Collect filtrate directly into a 2- to 3-m length of sterilized dialysis tubing that is attached to the funnel outlet and sealed at the opposite end. When filtration is complete, pour 100 mL Tricine buffer through the filter. Express residual fluid from the glass wool with a sterile, wooden tongue depressor and collect in dialysis bag.

c. Hydroextraction-dialysis concentration

Remove dialysis tube from filter outlet, seal open end, rinse bag exterior with sterile distilled water, and immediately place in a large beaker or pan. Add sufficient PEG-H_2O to completely cover all exterior surfaces of the dialysis bag. Hold container with its contents in the cold (4 to 5°C) overnight. Next morning, remove dialysis bag from PEG container and thoroughly rinse exterior with sterile distilled water. Open one end of bag, add 30 mL Tricine/$MgCl_2$/DEAE-dextran, and reclose. Thoroughly knead solution through the bag, either by hand or with a rubber spatula or a roller device to elute and resuspend viruses. Reopen one end of bag and, using a roller device or a rubber spatula, squeeze fluid into a 150-mL, sterile, screw cap centrifuge tube.

d. Final clarification and toxicity removal

Add 1 g diatomaceous earth, adjust to pH about 8.5 with 1.0 or $0.1N$ NaOH, and mix. If sample toxicity is expected, add ½ volume of trichlorotrifluoroethane and emulsify either by shaking the sealed tube vigorously for 1 min or by sonicating (at about 75 watts) for 30 s. If necessary chill sample in an ice bath during sonification to prevent overheating. Centrifuge at $2000 \times g$ for 30 min and collect aqueous supernatant by decanting or by aspirating with a pipet. Do not collect the lower organic layer. Add antibotics and store until virus assay (see Sections 5.2D and 5.2E).

5.4 ENTERIC VIRUS CONCENTRATION FROM SEAWATER AND ESTUARINE WATER

A. Introduction and General Description

The most widely used and successful methods for concentrating viruses from water involve four basic steps: virus adsorption to microporous filters under acid conditions, virus elution from filters with a small volume of alkaline solution, precipitation of viruses from the eluate, and recovery of the precipitated viruses by redissolving the precipitate in buffer.[18–22] A preferred method can not be specified.

The method described here is relatively simple (Figure 5:4).[21,23,24] The sample is acidified to pH 3.5 and $AlCl_3$ is added to enhance virus adsorption to the electronegative adsorbent filters. The conditioned water is filtered through two adsorbent cartridge filters in series: a fiberglass, yarn-wound prefilter followed by a microporous, pleated, thin-sheet, fiberglass-epoxy filter. From 100 to 400 liters are filtered, depending upon water quality. The filters are washed to remove excess aluminum ions and adsorbed viruses are recovered from the filters by filtering an elution fluid consisting of 1% beef extract-$0.05M$ glycine, pH 9.5. Eluted viruses are concentrated further by acidifying the eluate to pH 3.5 with HCl to acid precipitate viruses, the virus-containing precipitate is recovered by centrifugation and it is redissolved at pH 7.2 to 7.4 in alkaline disodium phosphate buffer.

The method is limited by suspended solids, dissolved inorganic salts, and dissolved organic matter in the water that may interfere with virus recovery or detection.[25] The procedure is not effective uniformly for all enteric viruses: some viruses present in field samples may not be detected.[18,26–28] Preferably make preliminary studies using samples of the water to be examined that have been seeded with representative enteric viruses.

B. Equipment, Materials and Reagents

1. *Equipment and materials*

Water pump: Centrifugal or positive displacement, with electrical or gas motor, capable of delivering 0.1 to 0.6 L/s at 207 kPa. If an electric pump is used, a gas-powered electric generator or other electricity source is needed. Sterilize pump head by recirculating a 10 mg free chlorine/L solution for 15 min and dechlorinate with a 50 mg $Na_2S_2O_3$/L solution for 3 min.

Adsorbent filters and housings: Prefilter: 25.4 cm-long, 3.0 μm

Laboratory Procedures for Seawater and Shellfish

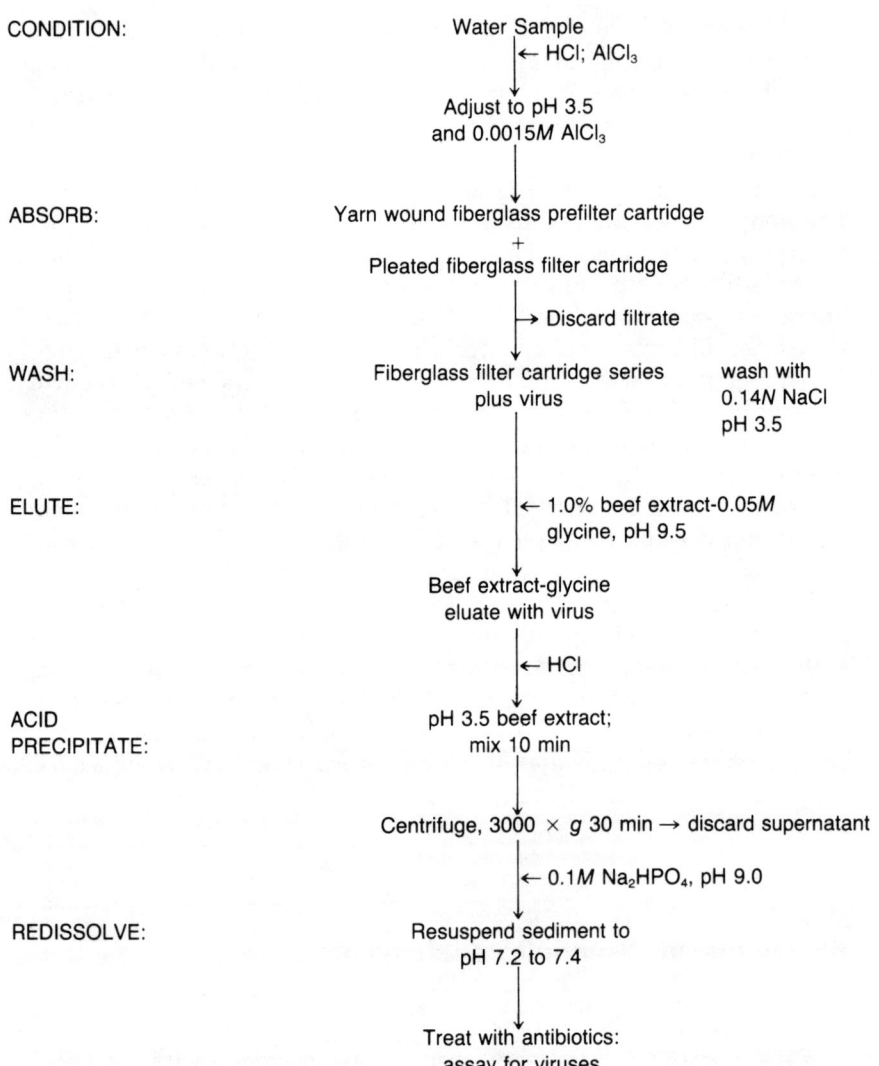

Figure 5:4—Concentration of enteric viruses from sea- and estuarine water by filter adsorption-elution and acid precipitation

nominal porosity, heat-treated fiberglass, yarn wound filter cartridge.*

Pleated filter: 25.4 cm-long, 0.45 μm nominal porosity, fiberglass-epoxy pleated filter cartridge.†

Filter housing: Autoclavable plastic housing with an optically clear or translucent bowl and a pressure relief valve in the head.**

pH meter: with combination electrode, battery operated for field use.

Pressure vessels: 4-L capacity, equipped with pressure gauge and inlet and outlet openings, and 12- to 20-L capacity, depending upon sample size, equipped with pressure gauge, inlet and outlet openings, and a manually operated precision metering valve on the outlet opening.††

Positive pressure source: up to 413 kPa, with regulator-laboratory air line, positive pressure pump, or cylinder of compressed air or nitrogen.

Hoses and connectors: nylon braid-reinforced, optically translucent plastic hose, ½-inch ID by ⅞-inch OD*** Fitted with stainless steel or brass connect/disconnect male fittings.††† Connect hoses to pumps, filter housings, and pressure vessels, and use to transfer water and other fluids during sample processing. Fit pumps, filter housings, pressure vessels, and other equipment to which hoses are to be connected with brass quick connect/disconnect female fittings.****

Sample container: for batch processing, sterilizable plastic or metal, 100 to 200 L capacity or larger, depending upon sample size.

Fluid proportioner: for continuous flow processing with automatic chemical feed system; with two, alternating feed pumps (duplex) and a mixing chamber.†††† Fit feed pump inlets with 4- to 6-

---

* No. K27R10S, Commercial Filter Div., Carborundum Co., Lebanon, Ind., or equivalent.

† No. DUO-FN 10-0.45, Duo-Fine, Filterite, Inc., Timonium, Md., or equivalent.

** No. LMO10UP-3/4, Filterite, Inc., Timonium, Md., or equivalent.

†† No. SS-4L2, Nupro Co., 4800 E. 345th St., Willoughby, Ohio, or equivalent.

*** No. HNB-50, Newage Corp., 2300 Maryland Rd., Willow Grove, Penn., or equivalent.

††† No. BPHN8-8, Snap-Tite, Inc., Union City, Penn., or equivalent.

**** No. BPHC8-8, Snap-Tite, Inc., Union City, Penn., or equivalent.

†††† Johansen and Son Machine Corp., Clifton, N.J., or equivalent.

foot long PVC hoses (3/8-inch ID × 1/2-inch OD). Set each chemical feed pump for a volume ratio of one part chemical feed to 200 parts water.

Two-way valve: for continuous flow processing with manual chemical feed system; use in-line between point of chemical feed into water line and prefilter inlet.

Tee connection: for continuous flow processing with manual chemical feed system; for adding chemical feed solution to sample in water line.

Pressure gauge: 0 to 700 kPa; mounted in a tee connection for in-line use between water pump and prefilter inlet.

Peristaltic or roller pump: for recirculation of elution fluid through adsorbent filters.

Centrifuge: capable of operating at 3000 × g; with rotors and buckets for 250- to 500-mL capacity bottles.

Magnetic stirrer: with stirring bars.

Centrifuge bottles: 250 to 500 mL.

Beakers: 2 and 4 L.

Graduated cylinders: 0.1, 1, and 2 L.

Pipets: 1, 5, and 10 mL.

Water flow meter: with brass quick connect/disconnect female fittings on the inlet and outlet. Connect to the outlet of pleated filter.

2. Reagents

Hydrochloric acid, HCl, 6.0, 1.0, 0.1, and 0.06N.

Sodium hydroxide, NaOH, 6, 1.0, and 0.1N.

Aluminum chloride, $AlCl_3$, 0.3 or 0.075M: Dissolve 72.4 or 18.1 g $AlCl_3 \cdot 6H_2O$/L distilled water. Autoclave.

Sodium chloride, NaCl, 0.14M, pH 3.5: Dissolve 8.18 g NaCl/L distilled water, autoclave, and aseptically adjust to pH 3.5 with 0.1N HCl.

1% beef extract-0.05M glycine, pH 9.5: Dissolve 10 g beef extract powder* and 3.75 g glycine/L distilled water. Adjust to pH 9.5 with 6N NaOH and autoclave.

Disodium phosphate, Na_2HPO_4, 0.1M, pH 9.0 to 9.5: Dissolve 26.8 g $Na_2HPO_4 \cdot 7H_2O$/L distilled water. Adjust to 9.0 to 9.5 with 1.0N NaOH if necessary and autoclave.

---

* Lablemco Powder, Oxoid USA, Inc., Columbia, Md., or equivalent.

Virological Examination

Sodium thiosulfate, $Na_2S_2O_3 \cdot 5\,H_2O$.
Sodium hypochlorite, 5.25% available chlorine (household bleach).
Antibiotics: See Section 5.2E.

C. Procedure

1. *Chemical addition and filter adsorption*

Use either a batch or continuous flow method to chemically condition the water sample and filter it through the virus adsorbent system.

    a. Batch method

        Connect sterile hoses to inlet and outlet of a sterilized water pump head. Place free end of inlet hose in water being sampled and free end of outlet hose in a sterile sample container. Operate pump and collect desired volume of sample (100 to 200 L or more). Place free ends of both pump hoses in sample container, making sure end of inlet hose is submerged. Place a pH electrode in the sample to monitor pH. Operate pump to mix sample by recirculation and *slowly* add 6N HCl until the pH is reduced to 3.8. Continue mixing and add 1 part 0.3M $AlCl_3$ per 200 parts (v/v) of water to give a final 0.0015M concentration of added $AlCl_3$. If sample is not at pH 3.5 ± 0.3, adjust by adding 6N HCl or NaOH, and turn off pump.

        Connect free end of pump outlet hose to inlet of prefilter, connect outlet of prefilter to inlet of pleated fiberglass filter, connect a hose to outlet of pleated fiberglass filter, and place free end of this filter outlet hose to discard discharged filtrate. Operate pump and filter entire sample at a flow rate of ≤ 1.3 L/s. Disconnect prefilter inlet from pump and connect prefilter inlet to a positive air (or nitrogen) pressure source. Operate pressure source to purge excess water from filters and filter housings, turn off pressure source, and disconnect from filters.

    b. Continuous flow methods

        i. Preparing HCl-$AlCl_3$ feed solution: Titrate 1 L of water to be sampled for viruses to pH 3.8 with 0.06N HCl and note volume used. Add 20 mL 0.075M $AlCl_3$ and, if necessary, add more 0.06N HCl to adjust pH to 3.5. Note volume of acid used: the total volume of HCl used determines milliliters of 1.0N HCl needed/L of 0.075M $AlCl_3$ to make the HCl-$AlCl_3$ feed solution. Make at least 3.0 L/100 L sample to be processed.

ii. Chemical addition and filtration: Use either a manually controlled system (Figure 5:5) or an hydraulically controlled system (Figure 5:6) to meter chemical feed solution during sample filtration.

1) Manual chemical feed system: Place feed solution in a sterile 12- to 20-L capacity pressure vessel equipped with a pressure gauge and a metering valve on the outlet. Using sterile hoses, connect pressure vessel inlet to a positive pressure source and the outlet to a tee connection in the water hose that connects pump outlet to adsorbent filter series inlet (Figure 5:5). Close metering valve on pressure vessel and set two-way valve in the water hose between the pump and filters to bypass position.

Operate pump to give a flow rate of ≤1.3 L/s. Note water line pressure and pressurize feed solution system to about 35 kPa above the water pressure. While checking pH of flowing water, slowly open feed system metering valve until water pH reaches 3.5 ± 0.3. Note water meter reading and turn two-way valve to direct water through the adsorbent filter series. Monitor the water meter and process desired total sample volume. Also monitor pH and, if necessary, adjust feed solution metering valve to maintain pH 3.5 ± 0.3. During processing, keep pressure of chemical feed system higher than water pressure. If chemical feed system pressure must be increased, do it slowly to avoid excess chemical feed. It may be necessary to decrease concurrently the chemical feed by adjusting the metering valve while slowly increasing chemical feed system pressure to keep pH in the acceptable range.

After filtering desired volume of water, turn two-way valve to bypass water flow. Close metering valve

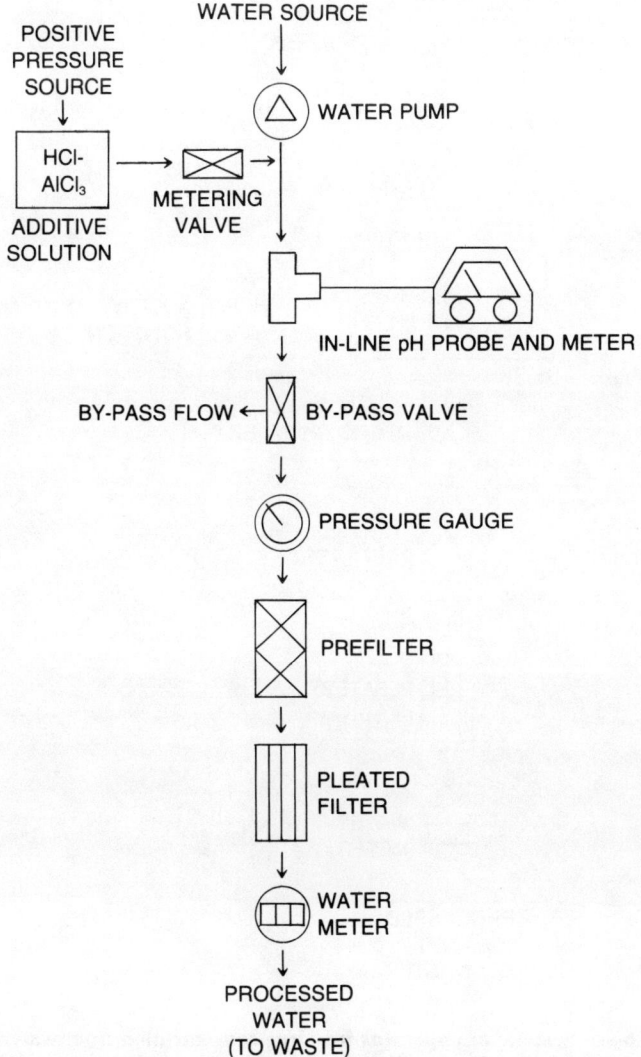

Figure 5:5—Schematic of apparatus for virus concentration from water using manually operated chemical feed system

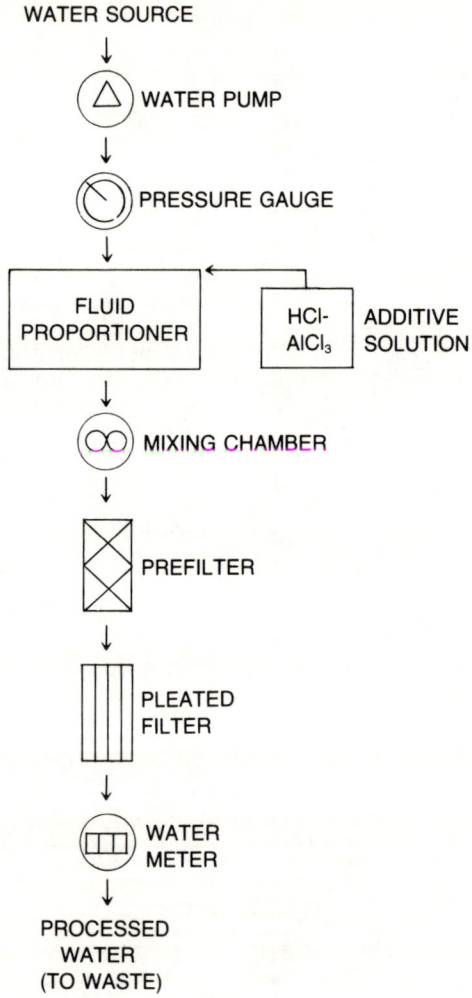

Figure 5:6—Schematic of apparatus for virus concentration from water using hydraulically operated chemical feed system

Virological Examination

of chemical feed system and depressurize. Turn off water pump and disconnect from filter system inlet. Connect prefilter inlet to a positive air (or nitrogen) pressure source and pressurize to purge excess water from filters and their housings. Turn off pressure source and disconnect from filters.

2) Hydraulically operated automatic chemical feed system: Place feed solution in a wide mouth container and drop free ends of the two hoses to the fluid proportioner feed pumps into the container so that they touch container bottom. Manually operate feed pump metering rods of the fluid proportioner to fill hoses and purge them of air. Place water pump inlet hose in the water source, connect water pump outlet hose to the fluid proportioner inlet, and connect a hose to fluid proportioner outlet. Turn on pump and briefly process water through fluid proportioner to purge it of excess chemical feed solution. Continue to process water, collect a small sample of conditioned water from the discharge of the fluid proportioner outlet hose, and measure pH. If pH is not 3.5 ± 0.3, adjust settings of both fluid proportioner feed pump rods. Operate proportioner again and recheck pH. Repeat adjustment until conditioned water pH is 3.5 ± 0.3.

Complete water processing assembly (Figure 5:6) by connecting fluid proportioner outlet hose to prefilter inlet. Connect prefilter outlet to pleated filter inlet. Connect water flow meter inlet to pleated filter outlet and connect a discharge hose to meter outlet. Note initial reading of meter. Add to this volume the desired volume of sample to be processed plus an additional 2% (to account for the volume of feed solution). Turn off water pump and stop sample processing at this meter reading. Turn on pump to begin processing water at a flow rate of 1.3 L/s. Immediately collect a filtrate sample from filter outlet hose and check pH. If pH is not 3.5 ± 0.3 turn off pump, disconnect fluid proportioner from prefilter inlet, and readjust fluid proportioner feed pump rods. Then operate water conditioning system and recheck pH. During sample processing, periodically check pH. When desired water volume is processed, turn off pump and disconnect fluid proportioner from prefilter inlet. Connect prefilter inlet to a positive air (or nitrogen) pressure source and pressurize to purge excess water from filters and their housings. Turn off pressure source and disconnect. Remove water meter from pleated filter outlet.

2. Saline wash of adsorbent filters

Place 4 L $0.15N$ NaCl, pH 3.5, in a pressure vessel. Connect pressure vessel inlet to a positive air (or nitrogen) pressure source and the outlet to the prefilter inlet. Slightly open vent (pressure relief) valves on the heads of the filter housings and slowly apply positive pressure to fill dead space in filter housings with wash solution. When solution just flows from the vents, quickly close them, and increase pressure to filter the solution through both filters in series. Discard filtrate. Let air (or nitrogen) purge excess fluid from filters and housings and depressurize the system.

3. Virus elution

Assemble elution system by connecting sterile hoses to prefilter inlet and pleated filter outlet. Place free end of both hoses in a 4-L beaker. Slip mid-section of prefilter inlet hose into the head of a peristaltic or roller pump. Add to 2 to 3 L elution medium (1% beef extract—$0.05M$ glycine, pH 9.5) to 4-L beaker. Open vents on the two filter housings slightly and slowly operate pump to fill the dead space in the filter housings with elution fluid. When fluid just flows from the vents, quickly close them and increase pump speed to recirculate elution fluid from beaker through filters for 5 min. Turn off pump and remove prefilter inlet hose from pump head. Connect free end of prefilter inlet hose to a positive pressure source and purge excess elution fluid from filters and their housings into beaker. Transfer sample to a sealable container and return to the laboratory under approved storage conditions for further processing.

4. Organic flocculation

Transfer eluate to a beaker and while mixing and measuring pH, slowly add $1N$ HCl to adjust to pH 3.5. Decrease mixing speed and mix slowly for 10 min for precipitate (floc) growth. Transfer to 250- to 500-mL capacity centrifuge bottles and centrifuge at 3000 × g for 10 min. Decant and discard the resulting supernatant. Dissolve sedimented precipitate in enough $0.1M$ Na_2HPO_4, pH 9.0 to 9.5, to bring the mixture to pH 7.2 to 7.4. Treat with antibiotics (Section 5.2E) and store for virus assay (Section 5.2D.).

5.5 ENTERIC VIRUS CONCENTRATION FROM SEDIMENTS

A. Introduction and General Description

The method described here is relatively simple and effective. It avoids the use of highly alkaline, non-proteinaceous elution fluids that rapidly inactivate some enteric viruses. In this method (Figure 5:7) the sediment is suspended in either 3% beef extract—$0.25M$ glycine, pH 9.5, or 3% beef extract—0.6% Tris buffer, pH 9.0 and is homogenized to enhance virus elution or extraction. The sediment particles are removed by centrifugation and viruses in the resulting supernatant fluid are concentrated by acid precipitation. The virus-containing precipitate is recovered by centrifugation and dissolved in a small volume of alkaline disodium phosphate.

This method for recovering viruses from sediments has been developed and evaluated using only a few different types of sea or estuarine sediments from specific geographic locations and only a few representative types of enteric viruses. Preliminary studies with seeded, representative virus types and the sediments to be examined are recommended to determine efficiency of virus recovery.

B. Equipment, Materials, and Reagents

1. *Equipment and materials*
    Sediment core sampler: 15 to 30 cm long by 2.5 to 10 cm diam with sterilizable head and autoclavable core tubes of plastic, glass, or corrosion-resistant metal.*
    Knife or spatula (sterilizable).

---

* No. 21WO188, Wards Natural Science Establishment, Inc., Rochester, N.Y., or similar device.

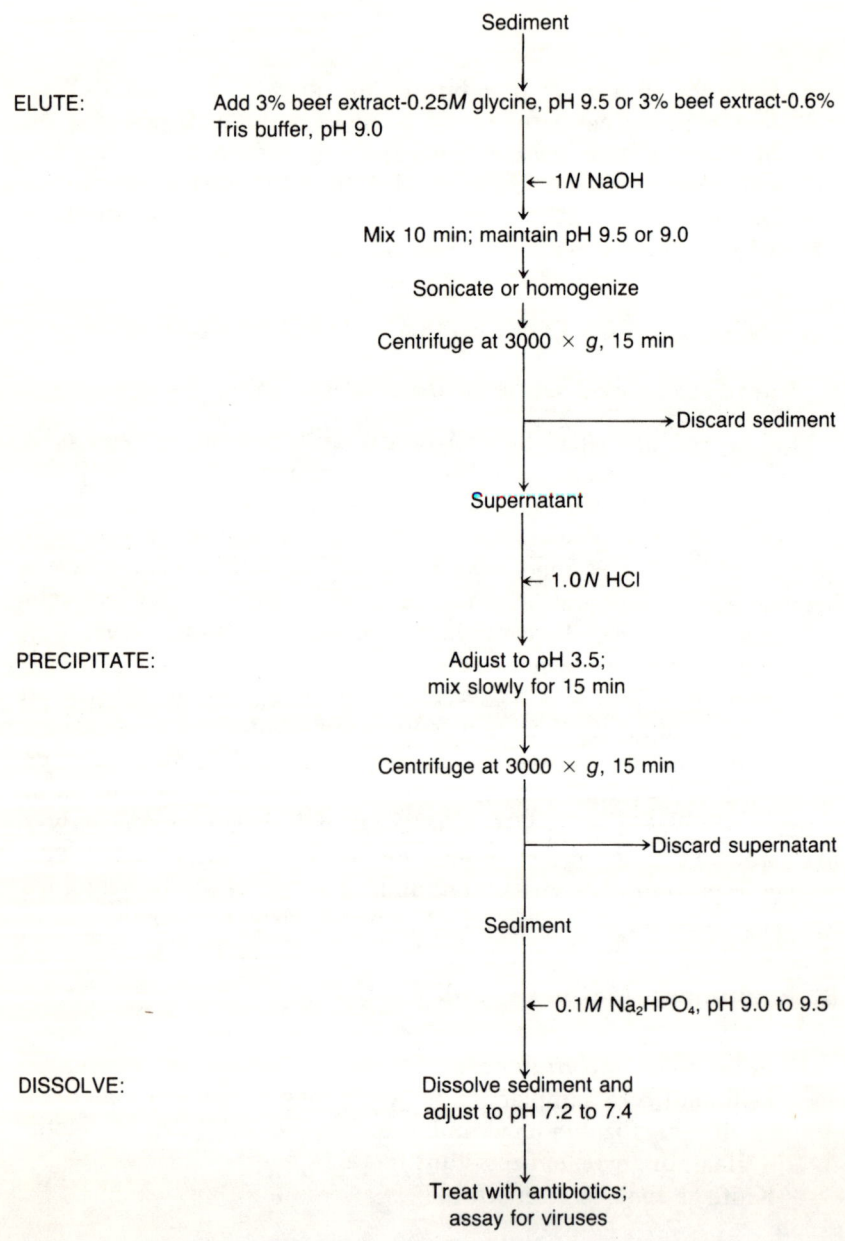

Figure 5:7—Concentration of enteric viruses from sediments by elution and acid precipitation

VIROLOGICAL EXAMINATION

Balance.
Beaker, 1 L.
Magnetic stirrer: with stirring bars.
Graduated cylinders, 0.1, 0.5, and 1.0 L.
Blender with sample container.
Sonicator (probe type optional).
pH meter with combination electrode.
Centrifuge: with rotors and buckets for 250- to 500-mL capacity bottles.
Centrifuge bottles: 250 to 500 mL.
Pipets, 1, 5, and 10 mL.

2. Reagents

Hydrochloric acid, HCl, 6.0 and 1.0N.
Sodium hydroxide, NaOH, 1.0N.
3% Beef extract-0.25M glycine, pH 9.5: Dissolve 30 g beef extract power* and 18.75 g glycine /L distilled water. Adjust to pH 9.5 with 6N HCl and autoclave.
3% Beef extract-0.6% Tris buffer, pH 9.0: Dissolve 30 g beef extract power* and 6 g Tris [(tris hydroxymethyl) aminomethane]/L sterile distilled water. Adjust to pH 9.5 with 1N NaOH or HCl and filter to sterilize.
Disodium phosphate, Na_2HPO_4, 0.1M, pH 9.0 to 9.5: Dissolve 26.8 g $Na_2HPO_4 \cdot 7H_2O$ per liter distilled water. Adjust to pH 9.0 to 9.5 if necessary and autoclave.
Antibiotics: See Section 5.2E.

C. Procedure

1. Sample collection and preparation

Because most viruses are likely to be in the first few centimeters of sediment, sample sediments using a relatively short core sampler. Collect several replicate cores from the same sampling site to minimize individual core variability in virus concentration and to obtain sufficient weight of sediment. Return intact cores to the

---

* Lablemco Powder, Oxoid USA, Inc. or equivalent.

laboratory and process as soon as possible, storing under approved conditions (Section 5.2D) until processed.

Aseptically remove cores from core tubes and divide into 2.5- to 7.5-cm long segments beginning at the top surface of the core. Most viruses will be found in the top segment, but deeper segments also may contain viruses and each can be processed separately. If necessary, composite corresponding core segments of two or more replicate samples from the same sampling site to obtain sufficient sediment weight.

2. Virus elution

Place 100 g or more sediment in a 1-L beaker, add 4 volumes (w/v) cold 3% beef extract-0.25M glycine, pH 9.5, (or 3% beef extract-0.6% Tris, pH 9.5), and mix vigorously for 10 min. Periodically during mixing, check pH and adjust to pH 9.5 with 1N NaOH if necessary. Chill and either sonicate at 100 watts for 5 min or homogenize in a blender for 1 to 2 min. Transfer to 250- to 500-mL capacity centrifuge bottles and centrifuge at 3000 × g for 10 min. Collect supernatant by decanting into a beaker.

3. Acid precipitation

While mixing sample and measuring its pH, slowly add 1N HCl to adjust to pH 3.5. Mix slowly for 10 min to promote precipitate (floc) growth. Transfer to 250- to 500-mL capacity centrifuge bottles and centrifuge at 3000 × g for 10 min. Decant and discard supernatant. Dissolve precipitate in enough 0.1M Na_2HPO_4, pH 9.0 to 9.5, to bring mixture to pH 7.2 to 7.4. Treat with antibiotics (Section 5.2E) and store until virus assay (Section 5.2D).

5.6 PROCEDURES FOR ISOLATING AND QUANTIFYING ENTERIC VIRUSES IN SHELLFISH, WATER, AND SEDIMENT CONCENTRATES

A. Introduction

Virus assays of concentrates from shellfish, water, and sediment samples are based on the infectivity of the viruses for susceptible live laboratory hosts. A complete description of facilities, equipment, materials, and methods for isolating enteric viruses from concentrated samples is beyond the scope of this chapter. Consult standard handbooks on virology methods for detailed information.[29,30] Virus assays are beyond the capacity of most shellfish sanitation laboratories. Such work should be done only by trained virologists working in specially equipped virology laboratory facilities. Take particular care to prevent samples or inoculated hosts from becoming contaminated with viruses from extraneous sources and to prevent cross-contamination of sample concentrates or inoculated hosts. In this section some important considerations in adapting standard virological methods to the examination of concentrate samples from shellfish, water, and sediment are described.

B. Host Systems for Virus Isolation and Assay

The major classes of host systems for detecting enteric virus infectivity are mammalian cell cultures of primate origin and suckling mice. Unfortunately, there is no single, universal host system that will detect all or even a majority of the viruses that potentially can contaminate shellfish and their habitat. Some of the epidemiologically most important viruses, such as hepatitis type A, Norwalk-type viruses, and human rotaviruses, cannot be conveniently isolated or assayed in any of the currently available, conventional host systems.

Primary or secondary human embryonic kidney and monkey kidney cell cultures and suckling mice are preferred host systems. However, few laboratories have the facilities and financial resources to maintain and utilize these host systems. A number of convenient and sensitive alternative host systems have been suggested.[31,32] Buffalo Green Monkey Kidney (BGM), a continuous line derived from African green monkey kidney cells is sensitive to a number of enteroviruses and reoviruses and is widely used as a substitute for primary monkey kidney cells for environmental samples. In addition, RD, a continuous line of human rhabdomyosarcoma cells, has been suggested as a substitute for suckling mice for detecting group A coxsackieviruses and also will detect a number of other enteric viruses.[32,33]

To obtain a wide spectrum of enteric viruses from shellfish, water, and sediment concentrates use at least two different host systems for virus isolation and assay.

C. Virus Assay Methods

1. *Quantal and enumerative assay procedures*

There are two basic procedures for virus isolation and assay in cell cultures: quantal methods ($TCID_{50}$ or MPN methods) and enumerative methods (plaque assay).[30] Quantal assays score replicate cultures of samples or sample dilutions as negative or positive for cytopathic effects (CPE). The virus titer is estimated from the resulting dose-response data. Plaque assays are considered more precise than quantal assays because relatively large numbers of individual infectious units are counted directly as discrete, localized areas of infection (plaques) in individual cultures. Plaque assay methods may use cell culture monolayers or cell culture suspensions in agar with the former most common for environmental samples.[33]

2. *Sample toxicity*

A major problem in virus isolation and assay of shellfish concentrates is toxicity of the samples to the cell cultures. Shellfish samples may contain toxic dissolved and particulate shellfish tissue components. Sample concentration schemes use various treatments intended to separate viruses from these toxic shellfish components but these treatments are not completely effective.

Shellfish toxicity may destroy all or most of the cells in the culture so that no virus isolations are obtained. Because of the possibility of "false-positive", non-viral cytopathology, confirm all CPE-positive cultures in quantal assays as virus-positive or virus-negative by further passage in cell cultures. An additional toxicity problem with plaque assays is the appearance of plaque-like cytotoxic areas in the cell cultures caused by localized, non-viral toxicants.[34,35] Because of such "false-positive", plaque-like areas of toxicity, confirm all or some representative proportion of the initial plaques for viruses by further passage.

When inoculating sample concentrates into cell cultures, take precautions to control toxicity. For example, the inoculum volume per cell culture and the adsorption time for the inoculum volume on the cell monolayer may influence sample toxicity. These conditions may have to be adjusted, depending on level of toxicity in sample concentrate. Draining the inoculum volume from monolayer cultures after the adsorption period and rinsing the monolayers with buffered saline, balanced salt solution, or cell culture medium also may be helpful in controlling sample toxicity.[36]

3. Mixed cultures

Another potential problem in quantal assays is the inoculation of two or more viruses into the same cell culture. If one virus grows faster than the other(s), the other virus(es) may not grow sufficiently to be detected. Alternatively, if two or more viruses grow, a mixed positive culture will be scored erroneously as a single positive, thus causing an underestimate of the number of viruses in the sample. Growth of two or more viruses in the culture will require separation of the individual virus types by further passage and isolation. Such mixed cultures may go undetected unless identification of the virus isolates is attempted. Mixed cultures are not a serious problem in quantal assays if the samples are divided into small portions and then inoculated into a series of replicate cultures[32] but the possibility of obtaining mixed cultures always must be considered.

4. Inoculation and adsorption conditions

The methods by which cell cultures are inoculated with and exposed to shellfish sample concentrates also may influence the efficiency of virus recovery. A widely used approach for monolayer cell cultures is to inoculate portions of the sample concentrate onto the monolayers of a series of cell cultures that have been drained of their growth or maintenance medium. The inoculum is left on the monolayers for a period of one to several hours to allow viruses to adsorb to the cells. During this time redistribute the inoculum over the monolayers every 15 to 20 min by tilting or rocking the cultures. After adsorption, liquid (for quantal assays) or agar-containing (for plaque assay) medium is added to the cultures. Alternatively, drain the inoculum from the cultures after the adsorption period and/or rinse the monolayers with buffered saline, balanced salt solution, or cell culture medium before adding liquid or agar-containing medium. Sometimes the inoculum is not removed and the cultures are not rinsed after adsorption.

When more than one type of cell culture is employed, most laboratories divide the entire sample into several equal portions and inoculate small volumes into a series of cultures of one type of cell (e.g. BGM or RD). A disadvantage of this approach is that the inoculated portions are exposed to only one type of cell. If viruses in the inoculum will not grow in that cell type, they will not be detected. This limitation can be overcome by passing material from cultures that remain virus-negative after incubation into fresh cell cultures of another type. Such "blind passage" of virus-negative cultures sometimes yields additional viruses, especially in quantal assays.

The inoculation and exposure conditions for shellfish concentrates in cell cultures recently has been investigated.[15] Optimum virus recoveries were obtained using inoculum volumes of ≤ 1 mL/45 cm^2 monolayer culture and an adsorption time of 2 h. Virus recoveries were increased by exposing the entire shellfish concentrate to each of two or more different cell culture types by means of a sequential adsorption scheme. In this sequential exposure scheme the inoculum volumes from initially inoculated cultures are recovered after the adsorption period, pooled, and then portions are reinoculated into another type of cell culture. The procedure can be repeated for inoculation of a third cell culture type.

5. *Virus isolation and identification*

In most laboratories inoculated cultures are incubated and periodically observed for the development of CPE or plaques for periods of one to three weeks or until the cells begin to degenerate as the cultures age. Material from CPE-positive cultures or plaques is confirmed for viruses by further passage into fresh cell cultures. Material from CPE-negative cultures also may be further passaged ("blind" passaged) because this sometimes yields additional virus isolates. Confirmed virus isolates can be identified using conventional serological techniques.[29,30]

5.7 REFERENCES

1. Ellender, R. D., J. B. Mapp, B. L. Middlebrooks, D. W. Cook & E. W. Cake. 1980. Natural enterovirus and fecal coliform contamination of Gulf Coast oysters. *J. Food Protect.* 43:105.
2. Gerba, C. P. & S. M. Goyal. 1978. Detection and occurrence of enteric viruses in shellfish: a review. *J. Food Protect.* 41:743.
3. Sobsey, M. D., C. R. Hackney, R. J. Carrick, B. Ray & M. L.

Speck. 1980. Occurrence of enteric bacteria and viruses in oysters. *J. Food Protect.* 43:111.
4. Vaughn, J. M., E. F. Landry, M. Z. Thomas, T. J. Vicale & W. F. Panello. 1980. Isolation of naturally occurring enteroviruses from a variety of shellfish species residing in Long Island and New Jersey marine embayments. *J. Food Protect.* 43:95.
5. Appleton, H. 1981. Outbreaks of viral gastroenteritis associated with shellfish. *In* M. Goddard & M. Butler, eds. Viruses and Wastewater Treatment, pp. 287–289, Pergamon Press, New York, N.Y.
6. Bostock, A. D., P. Mepham, S. Phillips, S. Skidmore & M. H. Hambling. 1979. Hepatitis A infection associated with the consumption of mussels. *J. Infect.* 1:171.
7. Dienstag, J. L., I. D. Gust, C. R. Lucas, D. C. Wong & R. H. Purcell. 1976. Mussel-associated viral hepatitis, type A: serological confirmation. *Lancet* i(7959):561.
8. Murphy, A. M., G. S. Grohmann, P. J. Christopher, et al. 1979. An Australia-wide outbreak of gastroenteritis from oysters caused by Norwalk virus. *Med. J. Aust.* 2:3290.
9. Portnoy, B. L., P. A. Mackowiak, C. T. Caraway, J. A. Walker, T. W. McKinley & C. A. Klein, Jr. 1975. Oyster associated hepatitis: Failure of shellfish certification programs to prevent outbreaks. *J. Amer. Med. Ass.* 233:1065.
10. Fugate, K. J., D. O. Cliver & M. T. Hatch. 1975. Enteroviruses and potential bacterial indicators in Gulf Coast oysters. *J. Milk and Food Technol.* 38:100.
11. Goyal, S. M., C. P. Gerba & J. L. Melnick. 1979. Human enteroviruses in oysters and their overlying waters. *Appl. Environ. Microbiol.* 37:572.
12. Sobsey, M. D., R. J. Carrick, B. E. Howard & H. R. Jensen. 1977. Improved methods for detecting enteric viruses in oysters and clams. *In:* Proceedings of the 10th National Shellfish Sanitation Workshop. D. S. Wilt, ed. Food and Drug Administration, U.S. Dept. of Health, Education and Welfare, Washington, D.C.
13. Sobsey, M. D., R. J. Carrick & J. R. Jensen. 1978. Improved methods for detecting enteric viruses in oysters. *Appl. Environ. Microbiol.* 36:121.
14. Cooper, R. C., K. M. Johnson, D. C. Straube, L. A. Brown & D. Lysmer. 1980. Development and evaluation of methods for the

detection of enteric viruses in San Francisco Bay shellfish, water and sediment. Report No. UCB/SERL 79-3, Sanitary Engineering Research Laboratory, University of California at Berkeley.
15. Metcalf, T. G., E. Moulton & D. Eckerson. 1980. Improved method and test strategy for recovery of enteric viruses from shellfish. *Appl. Environ. Microbiol.* 39:141.
16. Landry, E. F., J. M. Vaughn & T. J. Vicale. 1980. Modified procedure for extraction of poliovirus from naturally-infected oysters using Cat-Floc and beef extract. *J. Food Protect.* 43:91.
17. Tierney, J. T., A. Fassolitis, D. Van Donsel, V. C. Rao, R. Sullivan & E. P. Larkin. 1980. Glass wool-hydroextraction method for recovery of human enteroviruses from shellfish. *J. Food Protect.* 43:102.
18. Bitton, G., B. N. Feldberg & S. R. Farrah. 1979. Concentration of enteroviruses from seawater and tapwater by organic flocculation using non-fat dry milk and casein. *Water, Air and Soil Pollut.* 12:187.
19. Farrah, S. R., S. M. Goyal, D. P. Gerba, C. Wallis & J. L. Melnick. 1976. Concentration of enteroviruses from estuarine water. *Appl. Environ. Microbiol.* 33:1192.
20. Gerba, C. P., S. R. Farrah, S. M. Goyal, C. Wallis & J. L. Melnick. 1978. Concentration of enteroviruses from large volumes of tap water, treated sewage, and seawater. *Appl. Environ. Microbiol.* 35:540.
21. Payment, P., C. P. Gerba, C. Wallis & J. L. Melnick. 1976. Methods for concentrating viruses from large volumes of estuarine water on pleated membranes. *Water Res.* 10:893.
22. Sobsey, M. D., C. P. Gerba, C. Wallis & J. L. Melnick. 1977. Concentration of enteroviruses from large volumes of turbid estuary water. *Can. J. Microbiol.* 23:770.
23. Katzenelson, E., B. Fattal & T. Hostovesky. 1976. Organic flocculation: an efficient second-step concentration method for the detection of viruses in tap water. *Appl. Environ. Microbiol.* 32:638.
24. Landry, E. F., J. M. Vaughn, M. Z. Thomas & T. J. Vicale. 1978. Efficiency of beef extract for the recovery of poliovirus from wastewater effluents. *Appl. Environ. Microbiol.* 36:544.
25. Sobsey, M. D. 1982. Quality of currently available methodology

for monitoring viruses in the environment. *Environ. Internat.* 7:39.
26. Williams, F. P. Jr. & W. Jakubowski. 1978. Large volume virus concentration: evaluation of the organic flocculation method for elution/reconcentration. Abstracts of the Annual Meeting of the Amer. Soc. for Microbiol., p. 200, Amer. Soc. for Microbiol., Washington, D.C.
27. Morris, R. & W. M. Waite. 1980. Evaluation of procedures for recovery of viruses from water—I. Concentration systems. *Water Res.* 14:795.
28. Sobsey, M. D., J. S. Glass, R. R. Jacobs & W. A. Rutala. 1980. Modification of the tentative standard methods for improved virus recovery efficiency, *J. Amer. Water Works Ass.* 72:350.
29. Lennette, E. H. & N. J. Schmidt, eds. 1979. Diagnostic Procedures for Viral, Rickettsial and Chlamydial Infections, 5th ed. American Public Health Association, Washington, D.C.
30. American Public Health Association, American Water Works Association, Water Pollution Control Federation. 1981. Standard Methods for the Examination of Water and Wastewater, 15th ed. American Public Health Association, Washington, D.C.
31. Dahling, D. G., G. Berg & D. Berman. 1974. BGM: A continuous cell line more sensitive than primary rhesus and African green kidney cells for the recovery of viruses from water. *Health Lab. Sci.* 11:275.
32. Schmidt, N. J., H. H. Ho, J. L. Riggs & E. H. Lennette. 1978. Comparative sensitivity of various cell culture systems for isolation of viruses from wastewater and fecal samples. *Appl. Environ. Microbiol.* 36:480.
33. Morris, R. & W. M. Waite. 1980. Evaluation of procedures for recovery of viruses from water—II. Detection system. *Water Res.* 14:795.
34. Leong, L. Y. C., S. J. Barrett & R. R. Trussell. 1978. False-positives in testing of secondary sewage for enteric viruses. Abstracts of the Annual Meetings of the American Society for Microbiology, p. 200, American Society for Microbiology, Washington, D.C.
35. Fannin, K. E., S. H. Abid, J. J. Bertucci, J. M. Reed, S. C. Vana & C. Lue-Hing. 1978. Significance of reporting infectious virus

or plaque forming unit concentrations from environmental samples. Abstracts of the Annual Meeting of the American Society for Microbiology, p. 200, American Society for Microbiology, Washington, D.C.
36. Howard, B. E. 1980. The development and assessment of a method for concentrating enteric viruses in the hard shell clam *Mercenaria mercenaria*. Masters Thesis, Department of Environmental Sciences and Engineering, School of Public Health, University of North Carolina, Chapel Hill.

5.8. BIBLIOGRAPHY

GALTSOFF, P. S. 1964. The American oyster *Crassostrea virginica* Gmelin. U.S. Fish Wildlife Serv. *Fish. Bull.* 64:1.

METCALF, T. G. & W. C. STILES. 1965. The accumulation of enteric viruses by the oyster *Crassostrea virginica*. *J. Infec. Dis.* 115:68.

U.S. PUBLIC HEALTH SERVICE, National Shellfish Sanitation Program. 1965. *Manual of Operations*, Parts I and II., Pub. No. 33, U.S.P.H.S., Washington, D.C.

MITCHELL, J. R., M. W. PRESNELL, E. W. AKIN, J. M. CUMMINS & O. C. LIU. 1966. Accumulation and elimination of poliovirus by the eastern oyster. *Amer. J. Epidemiol.* 84:40.

HERMAN, J. E. & D. O. CLIVER. 1968. Methods for detecting foodborne enteroviruses. *Appl. Microbiol.* 16:1564.

KONOWALCHUK, J. & J. I. SPEIRS. 1972. Enterovirus recovery from laboratory-contaminated samples of shellfish. *Can. J. Microbiol.* 18:1023.

KOSTENBADER, K. D. & D. O. CLIVER. 1972. Polyelectrolyte flocculation as an aid to recovery of enteroviruses from oysters. *Appl. Microbiol.* 24:540.

HILL, W. F., JR., E. W. AKIN, W. BENTON, C. J. MAYHEW & T. G. METCALF. 1974. Recovery of poliovirus from turbid estuarine water on microporous filters by the use of Celite. *Appl. Microbiol.* 27:506.

METCALF, T. G., C. WALLIS & J. L. MELNICK. 1974. Virus enumeration and public health assessment in polluted surface water contributing to transmission of virus in nature. *In:* J. F. MALINA, JR. & B. P.

SAGIK, eds., *Virus Survival in Water and Wastewater Systems.* Center for Research in Water Resources, University of Texas, Austin.

METCALF, T. G., C. WALLIS & J. L. MELNICK. 1974. Environmental factors influencing isolation of enteroviruses from polluted surface waters. *Appl. Microbiol.* 27:920.

DEFLORA, S., G. P. DERENZI & G. BADOLATI. 1975. Detection of animal viruses in coastal seawater and sediments. *Appl. Microbiol.* 30:472.

SCHMIDT, N. J., H. H. HO & E. H. LENNETTE. 1975. Propagation and isolation of group A coxsackieviruses in RD cells. *J. Clin. Microbiol.* 2:183.

VAUGHN, J. M. & T. G. METCALF. 1975. Coliphages as indicators of enteric viruses in shellfish and shellfish-raising estuarine waters. *Water Res.* 9:613.

WILT, D. S. 1975. *Proceedings of the Ninth National Shellfish Sanitation Workshop.* Shellfish Sanitation Branch, Food and Drug Administration, U.S. Public Health Service, Washington, D.C.

GOLDFIELD, M. 1976. Epidemiological indicators for transmission of viruses in water. *In* BERG, G., H. L. BODILY, E. H. LENNETTE, M. L. MELNICK & T. G. METCALF, eds., *Viruses in Water.* American Public Health Association, Washington, D.C.

APPLETON, H. & M. PEREIRA. 1977. A possible virus aetiology in outbreaks of food poisoning from cockles. *Lancet* i(8015):780.

GERBA, C. P., E. M. SMITH & J. L. MELNICK. 1977. Development of a quantitative method for detecting enteroviruses in estuarine sediments. *Appl. Environ. Microbiol.* 34:158.

PAYMENT, P., M. TRUDEL & V. PAVILANIS. 1978. Evaluation de l'efficacite de la technique d'adsorption-elution du poliovirus 1 sur filtres en fibres de verre: application a l'analyse virologique de 100 ml a 100 L d'eau. *Can. J. Microbiol.* 24:1413.

SOBSEY, M. D., R. J. CARRICK & J. R. JENSEN. 1978. Improved methods for enteric viruses in oysters. *Appl. Environ. Microbiol.* 36:121.

LABELLE, R. L., C. P. GERBA, S. M. GOYAL, J. L. MELNICK, I. CECH & G. F. BOGDAN. 1980. Relationships between environmental factors, bacterial indicators, and the occurrence of enteric viruses in estuarine sediments. *Appl. Environ. Microbiol.* 39:588.

LARKIN, E. P. & T. G. METCALF. 1980. Cooperative study of methods for the recovery of enteric viruses from shellfish. *J. Food Protect.* 43:84.

SOBSEY, M. D., J. S. GLASS, R. J. CARRICK, R. R. JACOBS & W. A. RUTALA. 1980. Evaluation of the tentative standard method for enteric virus concentration from large volumes of tap water. *J. Amer. Water Works Ass.* 72:292.

APPENDIX

TABLE 1. DIFFERENCES TO CONVERT HYDROMETER READINGS AT ANY TEMPERATURE CENTIGRADE TO DENSITY AT 15°C

Observed Reading	Temperature of Water in Jar (°C)												
	-2.0	-1.0	0.0	1.0	2.0	3.0	4.0	5.0	6.0	7.0	8.0	9.0	10.0
0.9960													
0.9970													
0.9980													
0.9990	−1	−2	−3	−4	−5	−5	−6	−6	−6	−6	−6	−5	−5
1.0000	−2	−3	−4	−5	−5	−6	−6	−6	−6	−6	−6	−5	−5
1.0010	−3	−4	−4	−5	−6	−6	−6	−7	−7	−6	−6	−6	−5
1.0020	−3	−4	−5	−6	−6	−7	−7	−7	−7	−7	−6	−6	−5
1.0030	−4	−5	−6	−6	−7	−7	−7	−7	−7	−7	−6	−6	−5
1.0040	−4	−5	−6	−7	−7	−7	−8	−8	−7	−7	−7	−6	−6
1.0050	−5	−6	−6	−7	−8	−8	−8	−8	−8	−7	−7	−6	−6
1.0060	−6	−6	−7	−8	−8	−8	−8	−8	−8	−8	−7	−6	−6
1.0070	−6	−7	−8	−8	−8	−8	−8	−8	−8	−8	−7	−7	−6
1.0080	−7	−8	−8	−9	−9	−9	−9	−9	−8	−8	−7	−7	−6
1.0090	−7	−8	−9	−9	−9	−9	−9	−9	−9	−8	−8	−7	−6
1.0100	−8	−9	−9	−10	−10	−10	−10	−9	−9	−8	−8	−7	−6
1.0110	−9	−9	−10	−10	−10	−10	−10	−10	−9	−9	−8	−7	−6
1.0120	−9	−10	−10	−10	−10	−10	−10	−10	−10	−9	−8	−7	−7
1.0130	−10	−10	−11	−11	−11	−11	−11	−10	−10	−9	−8	−8	−7
1.0140	−10	−11	−11	−11	−11	−11	−11	−11	−10	−10	−9	−8	−7

1.0150	−11	−11	−12	−12	−12	−11	−11	−10	−10	−9	−8	−7
1.0160	−12	−12	−12	−12	−12	−12	−11	−11	−10	−9	−8	−7
1.0170	−12	−12	−13	−13	−13	−12	−12	−11	−10	−9	−8	−7
1.0180	−13	−13	−13	−13	−13	−13	−12	−11	−10	−9	−8	−7
1.0190	−13	−13	−14	−14	−13	−13	−12	−12	−11	−10	−9	−8
1.0200	−14	−14	−14	−14	−14	−13	−13	−12	−11	−10	−9	−8
1.0210	−14	−14	−14	−14	−14	−13	−13	−12	−11	−10	−9	−8
1.0220	−15	−15	−15	−15	−15	−14	−13	−12	−11	−10	−9	−8
1.0230	−15	−15	−15	−15	−15	−14	−13	−12	−11	−10	−9	−8
1.0240	−16	−16	−16	−16	−15	−14	−14	−13	−12	−11	−10	−8
1.0250	−16	−16	−16	−16	−16	−15	−14	−13	−12	−11	−10	−8
1.0260	−17	−17	−16	−16	−16	−15	−14	−13	−12	−11	−10	−8
1.0270	−18	−17	−17	−17	−17	−15	−14	−14	−12	−11	−10	−9
1.0280	−18	−18	−18	−17	−17	−16	−15	−14	−13	−11	−10	−9
1.0290	−19	−18	−18	−18	−17	−16	−15	−14	−13	−12	−10	−9
1.300	−19	−19	−19	−18	−18	−17	−16	−15	−14	−13	−12	−9
1.310	−20	−19	−19	−19	−18	−17	−16	−15	−14	−13	−12	−9

SOURCE: Zerbe, W. B. & C. B. Taylor. 1953. Sea Water Temperature and Density Reduction Tables. Coast and Geodetic Survey, U.S. Dept. of Commerce. Special Pub. No. 298, Washington, D.C.

(*continued*)

TABLE 1—(CONTINUED)

Observed Reading	Temperature of Water in Jar (°C)											
	11.0	12.0	13.0	14.0	15.0	16.0	17.0	18.0	18.5	19.0	19.5	20.0
0.9960												
0.9970												
0.9980							3	4	5	6	7	8
0.9990	−4	−3	−2	−1	0	1	3	4	5	6	7	8
1.0000	−4	−3	−2	−1	0	1	3	4	5	6	7	8
1.0010	−4	−3	−2	−1	0	1	3	4	5	6	7	8
1.0020	−4	−3	−2	−1	0	1	3	4	5	6	7	8
1.0030	−4	−3	−2	−1	0	1	3	4	5	6	7	8
1.0040	−5	−4	−3	−1	0	2	3	5	6	6	7	8
1.0050	−5	−4	−3	−1	0	2	3	5	6	7	8	9
1.0060	−5	−4	−3	−1	0	2	3	5	6	7	8	9
1.0070	−5	−4	−3	−2	0	2	3	5	6	7	8	9
1.0080	−5	−4	−3	−2	0	2	3	5	6	7	8	9
1.0090	−5	−4	−3	−2	0	2	3	5	6	7	8	9
1.0100	−5	−4	−3	−2	0	2	3	5	6	7	8	9
1.0110	−5	−4	−3	−2	0	2	3	5	6	7	8	9
1.0120	−6	−4	−3	−2	0	2	3	5	6	7	8	9
1.0130	−6	−4	−3	−2	0	2	4	5	6	7	8	10
1.0140	−6	−4	−3	−2	0	2	4	5	6	8	9	10

1.0150	−6	−4	−3	−2	0	2	4	5	6	8	9	10
1.0160	−6	−5	−3	−2	0	2	4	6	7	8	9	10
1.0170	−6	−5	−3	−2	0	2	4	6	7	8	9	10
1.0180	−6	−5	−3	−2	0	2	4	6	7	8	9	10
1.0190	−6	−5	−3	−2	0	2	4	6	7	8	9	10
1.0200	−6	−5	−3	−2	0	2	4	6	7	8	9	10
1.0210	−6	−5	−3	−2	0	2	4	6	7	8	9	10
1.0220	−7	−5	−3	−2	0	2	4	6	7	8	9	11
1.0230	−7	−5	−4	−2	0	2	4	6	7	8	9	11
1.0240	−7	−5	−4	−2	0	2	4	6	7	8	10	11
1.0250	−7	−5	−4	−2	0	2	4	6	7	8	10	11
1.0260	−7	−5	−4	−2	0	2	4	6	7	9	10	11
1.0270	−7	−5	−4	−2	0	2	4	6	7	9	10	11
1.0280	−7	−6	−4	−2	0	2	4	6	8	9	10	11
1.0290	−7	−6	−4	−2	0	2	4	6	8	9	10	11
1.0300	−7	−6	−4	−2	0	2	4	6	8	9	10	12
1.0310	−8	−6	−4	−1	0	2	4	6	8	9	10	12

(*continued*)

TABLE 1—(CONTINUED)

Observed Reading	Temperature of Water in Jar (°C)												
	20.5	21.0	21.5	22.0	22.5	23.0	23.5	24.0	24.5	25.0	25.5	26.0	26.5
0.9960													21
0.9970													22
0.9980	9	10	10	11	12	14	15	16	17	18	19	20	22
0.9990	9	10	11	12	13	14	15	16	17	18	19	20	22
1.0000	9	10	11	12	13	14	15	16	17	18	19	21	22
1.0010	9	10	11	12	13	14	15	17	18	19	20	21	23
1.0020	9	10	11	12	13	14	16	17	18	19	20	22	23
1.0030	9	10	11	12	13	15	16	17	18	19	21	22	23
1.0040	9	10	11	12	14	15	16	17	18	20	21	22	23
1.0050	10	11	12	13	14	15	16	17	19	20	21	22	24
1.0060	10	11	12	13	14	15	16	18	19	20	21	23	24
1.0070	10	11	12	13	14	15	17	18	19	20	21	23	24
1.0080	10	11	12	13	14	16	17	18	19	20	22	23	24
1.0090	10	11	12	13	15	16	17	18	19	21	22	23	25
1.0100	10	11	12	14	15	16	17	18	20	21	22	24	25
1.0110	10	12	13	14	15	16	17	19	20	21	22	24	25
1.0120	10	12	13	14	15	16	18	19	20	21	23	24	25
1.0130	11	12	13	14	15	16	18	19	20	22	23	24	26
1.0140	11	12	13	14	15	17	18	19	20	22	23	24	26

1.0150	11	12	13	14	16	17	18	20	21	22	23	25	26
1.0160	11	12	13	14	16	17	18	20	21	22	24	25	26
1.0170	11	12	13	15	16	17	18	20	21	22	24	25	27
1.0180	11	12	14	15	16	17	19	20	21	23	24	25	27
1.0190	11	12	14	15	16	18	19	20	21	23	24	26	27
1.0200	11	13	14	15	16	18	19	20	22	23	24	26	27
1.0210	12	13	14	15	17	18	19	21	22	23	25	26	27
1.0220	12	13	14	15	17	18	19	21	22	23	25	26	28
1.0230	12	13	14	16	17	18	20	21	22	24	25	26	28
1.0240	12	13	14	16	17	18	20	21	22	24	25	27	28
1.0250	12	13	15	16	17	18	20	21	23	24	25	27	28
1.0260	12	13	15	16	17	19	20	22	23	24	26	27	29
1.0270	12	14	15	16	18	19	20	22	23	24	26	27	29
1.0280	12	14	15	16	18	19	20	22	23	25	26	28	29
1.0290	13	14	15	16	18	19	21	22	23	25			
1.0300	13	14	15	16	18								
1.0310													

(continued)

TABLE 1—(CONCLUDED)

Observed Reading	Temperature of Water in Jar (°C)												
	27.0	27.5	28.0	28.5	29.0	29.5	30.0	30.5	31.0	31.5	32.0	32.5	33.0
0.9960	23	24	25	27	28	29	31	32	34	35	37	38	40
0.9970	23	24	26	27	28	30	31	33	34	36	37	39	40
0.9980	23	25	26	27	29	30	31	33	34	36	38	39	41
0.9990	24	25	26	28	29	30	32	33	35	36	38	39	41
1.0000	24	25	26	28	29	31	32	34	35	37	38	40	41
1.0010	24	25	27	28	30	31	32	34	35	37	39	40	42
1.0020	24	26	27	28	30	31	33	34	36	37	39	41	42
1.0030	25	26	27	29	30	32	33	35	36	38	39	41	42
1.0040	25	26	28	29	30	32	33	35	36	38	40	41	43
1.0050	25	26	28	29	31	32	34	35	37	38	40	42	43
1.0060	25	27	28	30	31	32	34	36	37	39	40	42	44
1.0070	26	27	28	30	31	33	34	36	38	39	41	42	44
1.0080	26	27	29	30	32	33	35	36	38	39	41	43	44
1.0090	26	28	29	30	32	33	35	36	38	40	41	43	45

Density													
1.0100	26	28	29	31	32	34	35	37	38	40	42	43	45
1.0110	27	28	30	31	32	34	36	37	39	40	42	44	45
1.0120	27	28	30	31	33	34	36	37	39	41	42	44	46
1.0130	27	29	30	32	33	35	36	38	39	41	43	44	46
1.0140	27	29	30	32	33	35	36	38	40	41	43	45	46
1.0150	28	29	31	32	34	35	37	38	40	42	43	45	47
1.0160	28	29	31	32	34	35	37	39	40	42	44	45	47
1.0170	28	30	31	33	34	36	37	39	40	42	44	46	47
1.0180	28	30	31	33	34	36	38	39	41	42	44	46	48
1.0190	29	30	32	33	35	36	38	39	41	43	44	46	48
1.0200	29	30	32	33	35	36	38	40	41	43	45	46	48
1.0210	29	31	32	34	35	37	38	40	42	43	45	47	49
1.0220	29	31	32	34	35	37	39	40	42	44	45	47	49
1.0230	30	31	33	34	36	37	39	41	42	44	46	47	49
1.0240	30	31	33	34	36	37	39	41	42	44	46	48	49
1.0250	30	31	33	35	36	38	39	41	43	44	46	48	50
1.0260	30	32	33	35	37	38	40	41	43	45	46	48	50
1.0270	30	32	34	35	37	38	40						
1.0280	31	32											
1.0290													
1.0300													
1.0310													

127

LABORATORY PROCEDURES FOR SEAWATER AND SHELLFISH

TABLE 2. CORRESPONDING DENSITIES AND SALINITIES*

Density	Salinity	Density	Salinity	Density	Salinity	Density	Salinity
0.9991	0.0	1.0026	4.5	1.0061	9.0	1.0096	13.6
0.9992	0.0	1.0027	4.6	1.0062	9.2	1.0097	13.7
0.9993	0.2	1.0028	4.7	1.0063	9.3	1.0098	13.9
0.9994	0.3	1.0029	4.8	1.0064	9.4	1.0099	14.0
0.9995	0.4	1.0030	5.0	1.0065	9.6	1.0100	14.1
0.9996	0.6	1.0031	5.1	1.0066	9.7	1.0101	14.2
0.9997	0.7	1.0032	5.2	1.0067	9.8	1.0102	14.4
0.9998	0.8	1.0033	5.4	1.0068	9.9	1.0103	14.5
0.9999	0.9	1.0034	5.5	1.0069	10.1	1.0104	14.6
1.0000	1.1	1.0035	5.6	1.0070	10.2	1.0105	14.8
1.0001	1.2	1.0036	5.8	1.0071	10.3	1.0106	14.9
1.0002	1.3	1.0037	5.9	1.0072	10.5	1.0107	15.0
1.0003	1.5	1.0038	6.0	1.0073	10.6	1.0108	15.2
1.0004	1.6	1.0039	6.2	1.0074	10.7	1.0109	15.3
1.0005	1.7	1.0040	6.3	1.0075	10.8	1.0110	15.4
1.0006	1.9	1.0041	6.4	1.0076	11.0	1.0111	15.6
1.0007	2.0	1.0042	6.6	1.0077	11.1	1.0112	15.7
1.0008	2.1	1.0043	6.7	1.0078	11.2	1.0113	15.8
1.0009	2.2	1.0044	6.8	1.0079	11.4	1.0114	16.0
1.0010	2.4	1.0045	6.9	1.0080	11.5	1.0115	16.1
1.0011	2.5	1.0046	7.1	1.0081	11.6	1.0116	16.2
1.0012	2.6	1.0047	7.2	1.0082	11.8	1.0117	16.3
1.0013	2.8	1.0048	7.3	1.0083	11.9	1.0118	16.5
1.0014	2.9	1.0049	7.5	1.0084	12.0	1.0119	16.6
1.0015	3.0	1.0050	7.6	1.0085	12.2	1.0120	16.7
1.0016	3.2	1.0051	7.7	1.0086	12.3	1.0121	16.9
1.0017	3.3	1.0052	7.9	1.0087	12.4	1.0122	17.0
1.0018	3.4	1.0053	8.0	1.0088	12.6	1.0123	17.1
1.0019	3.5	1.0054	8.1	1.0089	12.7	1.0124	17.3
1.0020	3.7	1.0055	8.2	1.0090	12.8	1.0125	17.4
1.0021	3.8	1.0056	8.4	1.0091	12.9	1.0126	17.5
1.0022	3.9	1.0057	8.5	1.0092	13.1	1.0127	17.7
1.0023	4.1	1.0058	8.6	1.0093	13.2	1.0128	17.8
1.0024	4.2	1.0059	8.8	1.0094	13.3	1.0129	17.9
1.0025	4.3	1.0060	8.9	1.0095	13.5	1.0130	18.0

* Density at 15°C. Salinity in parts per 1000 (‰, g/kg). (continued)

TABLE 2—(CONTINUED)

Density	Salinity	Density	Salinity	Density	Salinity	Density	Salinity
1.0131	18.2	1.0166	22.7	1.0201	27.3	1.0236	31.9
1.0132	18.3	1.0167	22.9	1.0202	27.5	1.0237	32.0
1.0133	18.4	1.0168	23.0	1.0203	27.6	1.0238	32.1
1.0134	18.6	1.0169	23.1	1.0204	27.7	1.0239	32.3
1.0135	18.7	1.0170	23.3	1.0205	27.8	1.0240	32.4
1.0136	18.8	1.0171	23.4	1.0206	28.0	1.0241	32.5
1.0137	19.0	1.0172	23.5	1.0207	28.1	1.0242	32.7
1.0138	19.1	1.0173	23.7	1.0208	28.2	1.0243	32.8
1.0139	19.2	1.0174	23.8	1.0209	28.4	1.0244	32.9
1.0140	19.3	1.0175	23.9	1.0210	28.5	1.0245	33.1
1.0141	19.5	1.0176	24.1	1.0211	28.6	1.0246	33.2
1.0142	19.6	1.0177	24.2	1.0212	28.8	1.0247	33.3
1.0143	19.7	1.0178	24.3	1.0213	28.9	1.0248	33.5
1.0144	19.9	1.0179	24.4	1.0214	29.0	1.0249	33.6
1.0145	20.0	1.0180	24.6	1.0215	29.1	1.0250	33.7
1.0146	20.1	1.0181	24.7	1.0216	29.3	1.0251	33.8
1.0147	20.3	1.0182	24.8	1.0217	29.4	1.0252	34.0
1.0148	20.4	1.0183	25.0	1.0218	29.5	1.0253	34.1
1.0149	20.5	1.0184	25.1	1.0219	29.7	1.0254	34.2
1.0150	20.6	1.0185	25.2	1.0220	29.8	1.0255	34.4
1.0151	20.8	1.0186	25.4	1.0221	29.9	1.0256	34.5
1.0152	20.9	1.0187	25.5	1.0222	30.1	1.0257	34.6
1.0153	21.0	1.0188	25.6	1.0223	30.2	1.0258	34.8
1.0154	21.2	1.0189	25.8	1.0224	30.3	1.0259	34.9
1.0155	21.3	1.0190	25.9	1.0225	30.4	1.0260	35.0
1.0156	21.4	1.0191	26.0	1.0226	30.6	1.0261	35.1
1.0157	21.6	1.0192	26.1	1.0227	30.7	1.0262	35.3
1.0158	21.7	1.0193	26.3	1.0218	30.8	1.0263	35.4
1.0159	21.8	1.0194	26.4	1.0229	31.0	1.0264	35.5
1.0160	22.0	1.0195	26.5	1.0230	31.1	1.0265	35.7
1.0161	22.1	1.0196	26.7	1.0231	31.2	1.0266	35.8
1.0162	22.2	1.0197	26.8	1.0232	31.4	1.0267	35.9
1.0163	22.4	1.0198	26.9	1.0233	31.5	1.0268	36.0
1.0164	22.5	1.0199	27.1	1.0234	31.6	1.0269	36.2
1.0165	22.6	1.0200	27.2	1.0235	31.8	1.0270	36.3

TABLE 2—(CONCLUDED)

Density	Salinity	Density	Salinity	Density	Salinity	Density	Salinity
1.0271	36.4	1.0286	38.4	1.0301	40.3	1.0316	42.3
1.0272	36.6	1.0287	38.5	1.0302	40.4	1.0317	42.4
1.0273	36.7	1.0288	38.6	1.0303	40.6	1.0318	42.5
1.0274	36.8	1.0289	38.8	1.0304	40.7	1.0319	42.7
1.0275	37.0	1.0290	38.9	1.0305	40.8	1.0320	42.8
1.0276	37.1	1.0291	39.0	1.0306	41.0		
1.0277	37.2	1.0292	39.2	1.0307	41.1		
1.0278	37.3	1.0293	39.3	1.0308	41.2		
1.0279	37.5	1.0294	39.4	1.0309	41.4		
1.0280	37.6	1.0295	39.6	1.0310	41.5		
1.0281	37.7	1.0296	39.7	1.0311	41.6		
1.0282	37.9	1.0297	39.8	1.0312	41.7		
1.0283	38.0	1.0298	39.9	1.0313	41.9		
1.0284	38.1	1.0299	40.1	1.0314	42.0		
1.0285	38.2	1.0300	40.2	1.0315	42.1		

TABLE 3. CONVERSION OF CHLOROSITY TO SALINITY

Conversion of 20°C chlorosity, $Cl/\text{liter}_{(20)}$, to salinity, $S‰$ from the expression
$$S‰ = 0.03 + [1.8050 \times Cl/\text{liter}_{(20)} \times 1\ \rho_{(20)}]$$
where $\rho_{(20)}$ is the density of seawater at chlorosity $Cl/\text{liter}_{(20)}$.

$Cl/\text{liter}_{(20)}$	$S‰$	$Cl/\text{liter}_{(20)}$	$S‰$	$Cl/\text{liter}_{(20)}$	$S‰$	$Cl/\text{liter}_{(20)}$	$S‰$
2.00	3.64	.35	.27	2.70	4.89	.05	.52
.01	.66	.36	.29	.71	.91	.06	.54
.02	.68	.37	.30	.72	.93	.07	.56
.03	.69	.38	.32	.73	.95	.08	.57
.04	.71	.39	.34	.74	.97	.09	.59
.05	.73	2.40	4.36	.75	4.98	3.10	5.61
.06	.75	.41	.37	.76	5.00	.11	.63
.07	.77	.42	.39	.77	.02	.12	.65
.08	.78	.43	.41	.78	.04	.13	.66
.09	.80	.44	.43	.79	.06	.14	.68
2.10	3.82	.45	.45	2.80	5.07	.15	.70
.11	.84	.46	.46	.81	.09	.16	.72
.12	.86	.47	.48	.82	.11	.17	.74
.13	.87	.48	.50	.83	.13	.18	.75
.14	.89	.49	.52	.84	.14	.19	.77
.15	.91	2.50	4.54	.85	.16	3.20	5.79
.16	.93	.51	.55	.86	.18	.21	.81
.17	.95	.52	.57	.87	.20	.22	.82
.18	.96	.53	.59	.88	.22	.23	.84
.19	3.98	.54	.61	.89	.24	.24	.86
2.20	4.00	.55	.63	2.90	5.25	.25	.88
.21	.02	.56	.64	.91	.27	.26	.90
.22	.03	.57	.66	.92	.29	.27	.91
.23	.05	.58	.68	.93	.31	.28	.93
.24	.07	.59	.70	.94	.32	.29	.95
.25	.09	2.60	4.71	.95	.34	3.30	5.97
.26	.11	.61	.73	.96	.36	.31	5.99
.27	.12	.62	.75	.97	.38	.32	6.00
.28	.14	.63	.77	.98	.40	.33	.02
.29	.16	.64	.79	.99	.41	.34	.04
2.30	4.18	.65	.80	3.00	5.43	.35	.06
.31	.20	.66	.82	.01	.45	.36	.08
.32	.21	.67	.84	.02	.47	.37	.09
.33	.23	.68	.86	.03	.48	.38	.11
.34	.25	.69	.88	.04	.50	.39	.13

SOURCE: Reproduced with permission from Strickland, J. O. H. & T. R. Parsons. 1965. *A Manual of Sea Water Analysis*. Director General of Printing and Publishing, Department of Public Printing and Stationery, Ottawa, Canada. (*continued*)

TABLE 3—(CONTINUED)

Cl/liter$_{(20)}$	$S‰$	Cl/liter$_{(20)}$	$S‰$	Cl/liter$_{(20)}$	$S‰$	Cl/liter$_{(20)}$	$S‰$
3.40	6.15	3.80	6.86	4.20	7.58	4.60	8.29
.41	.16	.81	.88	.21	.60	.61	.31
.42	.18	.82	.90	.22	.61	.62	.33
.43	.20	.83	.92	.23	.63	.63	.35
.44	.22	.84	.93	.24	.65	.64	.36
.45	.24	.85	.95	.25	.67	.65	.38
.46	.25	.86	.97	.26	.68	.66	.40
.47	.27	.87	6.98	.27	.70	.67	.42
.48	.29	.88	7.01	.28	.72	.68	.44
.49	.31	.89	.02	.29	.74	.69	.45
3.50	6.33	3.90	7.04	4.30	7.76	4.70	8.47
.51	.34	.91	.06	.31	.77	.71	.49
.52	.36	.92	.08	.32	.79	.72	.51
.53	.38	.93	.10	.33	.81	.73	.52
.54	.40	.94	.11	.34	.83	.74	.54
.55	.42	.95	.13	.35	.85	.75	.56
.56	.43	.96	.15	.36	.86	.76	.58
.57	.45	.97	.17	.37	.88	.77	.60
.58	.47	.98	.18	.38	.90	.78	.61
.59	.49	.99	.20	.39	.92	.79	.63
3.60	6.50	4.00	7.22	4.40	7.93	4.80	8.65
.61	.52	.01	.24	.41	.95	.81	.67
.62	.54	.02	.26	.42	.97	.82	.69
.63	.56	.03	.27	.43	7.99	.83	.70
.64	.58	.04	.29	.44	8.01	.84	.72
.65	.59	.05	.31	.45	.02	.85	.74
.66	.61	.06	.33	.46	.04	.86	.76
.67	.63	.07	.35	.47	.06	.87	.77
.68	.65	.08	.36	.48	.08	.88	.79
.69	.67	.09	.38	.49	.10	.89	.81
3.70	6.68	4.10	7.40	4.50	8.11	4.90	8.83
.71	.70	.11	.42	.51	.13	.91	.85
.72	.72	.12	.43	.52	.15	.92	.86
.73	.74	.13	.45	.53	.17	.93	.88
.74	.76	.14	.47	.54	.18	.94	.90
.75	.77	.15	.49	.55	.20	.95	.92
.76	.79	.16	.51	.56	.22	.96	.94
.77	.81	.17	.52	.57	.24	.97	.95
.78	.83	.18	.54	.58	.26	.98	.97
.79	.84	.19	.56	.59	.27	.99	.99

TABLE 3—(CONTINUED)

Cl/liter$_{(20)}$	S‰	Cl/liter$_{(20)}$	S‰	Cl/liter$_{(20)}$	S‰	Cl/liter$_{(20)}$	S‰
5.00	9.01	5.40	9.72	5.80	10.43	6.20	11.14
.01	.02	.41	.74	.81	.45	.21	.16
.02	.04	.42	.75	.82	.47	.22	.18
.03	.06	.43	.77	.83	.48	.23	.20
.04	.08	.44	.79	.84	.50	.24	.21
.05	.10	.45	.81	.85	.52	.25	.23
.06	.11	.46	.83	.86	.54	.26	.25
.07	.13	.47	.84	.87	.56	.27	.27
.08	.15	.48	.86	.88	.57	.28	.28
.09	.17	.49	.88	.89	.59	.29	.30
5.10	9.18	5.50	9.90	5.90	10.61	6.30	11.32
.11	.20	.51	.91	.91	.63	.31	.34
.12	.22	.52	.93	.92	.64	.32	.36
.13	.24	.53	.95	.93	.66	.33	.37
.14	.26	.54	.97	.94	.68	.34	.39
.15	.27	.55	.99	.95	.70	.35	.41
.16	.29	.56	10.00	.96	.72	.36	.43
.17	.31	.57	.02	.97	.73	.37	.44
.18	.33	.58	.04	.98	.75	.38	.46
.19	.34	.59	.06	.99	.77	.39	.48
5.20	9.36	5.60	10.07	6.00	10.79	6.40	11.50
.21	.38	.61	.09	.01	.81	.41	.52
.22	.40	.62	.11	.02	.82	.42	.53
.23	.42	.63	.13	.03	.84	.43	.55
.24	.43	.64	.15	.04	.86	.44	.57
.25	.45	.65	.16	.05	.88	.45	.59
.26	.47	.66	.18	.06	.89	.46	.60
.27	.49	.67	.20	.07	.91	.47	.62
.28	.50	.68	.22	.08	.93	.48	.64
.29	.52	.69	.24	.09	.95	.49	.66
5.30	9.54	5.70	10.25	6.10	10.97	6.50	11.68
.31	.56	.71	.27	.11	10.98	.51	.69
.32	.58	.72	.29	.12	11.00	.52	.71
.33	.59	.73	.31	.13	.02	.53	.73
.34	.61	.74	.32	.14	.04	.54	.75
.35	.63	.75	.34	.15	.05	.55	.76
.36	.65	.76	.36	.16	.07	.56	.78
.37	.67	.77	.38	.17	.09	.57	.80
.38	.68	.78	.40	.18	.11	.58	.82
.39	.70	.79	.41	.19	.12	.59	.84

Table 3—(Continued)

Cl/liter$_{(20)}$	$S‰$	Cl/liter$_{(20)}$	$S‰$	Cl/liter$_{(20)}$	$S‰$	Cl/liter$_{(20)}$	$S‰$
6.60	11.85	7.00	12.56	7.40	13.27	7.80	13.98
.61	.87	.01	.58	.41	.29	.81	14.00
.62	.89	.02	.60	.42	.31	.82	.02
.63	.91	.03	.62	.43	.33	.83	.03
.64	.92	.04	.63	.44	.34	.84	.05
.65	.94	.05	.65	.45	.36	.85	.07
.66	.96	.06	.67	.46	.38	.86	.09
.67	11.98	.07	.69	.47	.40	.87	11
.68	12.00	.08	.71	.48	.41	.88	12
.69	.01	.09	.72	.49	.43	.89	.14
6.70	12.03	7.10	12.74	7.50	13.45	7.90	14.16
.71	.05	.11	.76	.51	.47	.91	.18
.72	.07	.12	.78	.52	.49	.92	.19
.73	.08	.13	.79	.53	.50	.93	.21
.74	.10	.14	.81	.54	.52	.94	.23
.75	.12	.15	.83	.55	.54	.95	.25
.76	.14	.16	.85	.56	.56	.96	.27
.77	.16	.17	.86	.57	.57	.97	.28
.78	.17	.18	.88	.58	.59	.98	.30
.79	.19	.19	.90	.59	.61	.99	.32
6.80	12.21	7.20	12.92	7.60	13.63	8.00	14.34
.81	.23	.21	.94	.61	.65	.01	.35
.82	.24	.22	.95	.62	.66	.02	.37
.83	.26	.23	.97	.63	.68	.03	.39
.84	.28	.24	12.99	.64	.70	.04	.41
.85	.30	.25	13.01	.65	.72	.05	.42
.86	.31	.26	.02	.66	.73	.06	.44
.87	.33	.27	.04	.67	.75	.07	.46
.88	.35	.28	.06	.68	.77	.08	.48
.89	.37	.29	.08	.69	.79	.09	.50
6.90	12.39	7.30	13.10	7.70	13.80	8.10	14.51
.91	.40	.31	.11	.71	.82	.11	.53
.92	.42	.32	.13	.72	.84	.12	.55
.93	.44	.33	.15	.73	.86	.13	.57
.94	.46	.34	.17	.74	.88	.14	.58
.95	.47	.35	.18	.75	.89	.15	.60
.96	.49	.36	.20	.76	.91	.16	.62
.97	.51	.37	.22	.77	.93	.17	.64
.98	.53	.38	.24	.78	.95	.18	.65
.99	.55	.39	.25	.79	.96	.19	.67

Table 3—(Continued)

Cl/liter$_{(20)}$	S‰	Cl/liter$_{(20)}$	S‰	Cl/liter$_{(20)}$	S‰	Cl/liter$_{(20)}$	S‰
8.20	14.69	8.60	15.40	9.00	16.10	9.40	16.81
.21	.71	.61	.41	.01	.12	.41	.82
.22	.72	.62	.43	.02	.14	.42	.84
.23	.74	.63	.45	.03	.16	.43	.86
.24	.76	.64	.47	.04	.17	.44	.88
.25	.78	.65	.48	.05	.19	.45	.89
.26	.80	.66	.50	.06	.21	.46	.91
.27	.81	.67	.52	.07	.23	.47	.93
.28	.83	.68	.54	.08	.24	.48	.95
.29	.85	.69	.56	.09	.26	.49	.96
8.30	14.87	8.70	15.57	9.10	16.28	9.50	16.98
.31	.88	.71	.59	.11	.30	.51	17.00
.32	.90	.72	.61	.12	.31	.52	.02
.33	.92	.73	.63	.13	.33	.53	.03
.34	.94	.74	.64	.14	.35	.54	.05
.35	.95	.75	.66	.15	.37	.55	.07
.36	.97	.76	.68	.16	.38	.56	.09
.37	14.99	.77	.70	.17	.40	.57	.11
.38	15.01	.78	.71	.18	.42	.58	.12
.39	.03	.79	.73	.19	.44	.59	.14
8.40	15.04	8.80	15.75	9.20	16.45	9.60	17.16
.41	.06	.81	.77	.21	.47	.61	.18
.42	.08	.82	.79	.22	.49	.62	.19
.43	.10	.83	.80	.23	.51	.63	.21
.44	.11	.84	.82	.24	.53	.64	.23
.45	.13	.85	.84	.25	.54	.65	.25
.46	.15	.86	.86	.26	.56	.66	.26
.47	.17	.87	.87	.27	.58	.67	.28
.48	.18	.88	.89	.28	.60	.68	.30
.49	.20	.89	.91	.29	.61	.69	.32
8.50	15.22	8.90	15.93	9.30	16.63	9.70	17.33
.51	.24	.91	.94	.31	.65	.71	.35
.52	.25	.92	.96	.32	.67	.72	.37
.53	.27	.93	15.98	.33	.68	.73	.39
.54	.29	.94	16.00	.34	.70	.74	.40
.55	.31	.95	.01	.35	.72	.75	.42
.56	.33	.96	.03	.36	.74	.76	.44
.57	.34	.97	.05	.37	.75	.77	.46
.58	.36	.98	.07	.38	.77	.78	.47
.59	.38	.99	.09	.39	.79	.79	.49

Table 3—(Continued)

Cl/liter$_{(20)}$	$S‰$	Cl/liter$_{(20)}$	$S‰$	Cl/liter$_{(20)}$	$S‰$	Cl/liter$_{(20)}$	$S‰$
9.80	17.51	10.20	18.22	10.60	18.92	11.00	19.62
.81	.53	.21	.23	.61	.94	.01	.64
.82	.54	.22	.25	.62	.96	.02	.66
.83	.56	.23	.27	.63	.97	.03	.68
.84	.58	.24	.29	.64	18.99	.04	.69
.85	.60	.25	.30	.65	19.01	.05	.71
.86	.62	.26	.32	.66	.03	.06	.73
.87	.63	.27	.34	.67	.04	.07	.75
.88	.65	.28	.36	.68	.06	.08	.76
.89	.67	.29	.38	.69	.08	.09	.78
9.90	17.69	10.30	18.39	10.70	19.10	11.10	19.80
.91	.70	.31	.41	.71	.11	.11	.82
.92	.72	.32	.43	.72	.13	.12	.83
.93	.74	.33	.45	.73	.15	.13	.85
.94	.76	.34	.46	.74	.17	.14	.87
.95	.77	.35	.48	.75	.18	.15	.89
.96	.79	.36	.50	.76	.20	.16	.90
.97	.81	.37	.52	.77	.22	.17	.92
.98	.83	.38	.53	.78	.24	.18	.94
.99	.85	.39	.55	.79	.25	.19	.96
10.00	17.87	10.40	18.57	10.80	19.27	11.20	19.97
.01	.88	.41	.59	.81	.29	.21	19.99
.02	.90	.42	.60	.82	.31	.22	20.01
.03	.92	.43	.62	.83	.32	.23	.03
.04	.94	.44	.64	.84	.34	.24	.04
.05	.95	.45	.66	.85	.36	.25	.06
.06	.97	.46	.67	.86	.38	.26	.08
.07	17.99	.47	.69	.87	.39	.27	.10
.08	18.01	.48	.71	.88	.41	.28	.11
.09	.02	.49	.73	.89	.43	.29	.13
10.10	18.04	10.50	18.74	10.90	19.45	11.30	20.15
.11	.06	.51	.76	.91	.47	.31	.17
.12	.08	.52	.78	.92	.48	.32	.18
.13	.09	.53	.80	.93	.50	.33	.20
.14	.11	.54	.81	.94	.52	.34	.22
.15	.13	.55	.83	.95	.54	.35	.24
.16	.15	.56	.85	.96	.55	.36	.26
.17	.16	.57	.87	.97	.57	.37	.27
.18	.18	.58	.88	.98	.59	.38	.29
.19	.20	.59	.90	.99	.61	.39	.31

Table 3—(Continued)

$Cl/\text{liter}_{(20)}$	$S‰$	$Cl/\text{liter}_{(20)}$	$S‰$	$Cl/\text{liter}_{(20)}$	$S‰$	$Cl/\text{liter}_{(20)}$	$S‰$
11.40	20.33	11.80	21.03	12.20	21.73	12.60	22.43
.41	.34	.81	.04	.21	.75	.61	.44
.42	.36	.82	.06	.22	.76	.62	.46
.43	.38	.83	.08	.23	.78	.63	.48
.44	.40	.84	.10	.24	.80	.64	.50
.45	.41	.85	.11	.25	.82	.65	.51
.46	.43	.86	.13	.26	.83	.66	.53
.47	.45	.87	.15	.27	.85	.67	.55
.48	.47	.88	.17	.28	.87	.68	.57
.49	.48	.89	.18	.29	.89	.69	.58
11.50	20.50	11.90	21.20	12.30	21.90	12.70	22.60
.51	.52	.91	.22	.31	.92	.71	.62
.52	.54	.92	.24	.32	.94	.72	.64
.53	.55	.93	.26	.33	.96	.73	.65
.54	.57	.94	.27	.34	.97	.74	.67
.55	.59	.95	.29	.35	21.99	.75	.69
.56	.61	.96	.31	.36	22.01	.76	.71
.57	.62	.97	.33	.37	.03	.77	.72
.58	.64	.98	.34	.38	.04	.78	.74
.59	.66	.99	.36	.39	.06	.79	.76
11.60	20.68	12.00	21.38	12.40	22.08	12.80	22.78
.61	.69	.01	.40	.41	.09	.81	.79
.62	.71	.02	.41	.42	.11	.82	.81
.63	.73	.03	.43	.43	.13	.83	.83
.64	.75	.04	.45	.44	.15	.84	.85
.65	.76	.05	.47	.45	.16	.85	.86
.66	.78	.06	.48	.46	.18	.86	.88
.67	.80	.07	.50	.47	.20	.87	.90
.68	.82	.08	.52	.48	.22	.88	.92
.69	.83	.09	.54	.49	.23	.89	.93
11.70	20.85	12.10	21.55	12.50	22.25	12.90	22.95
.71	.87	.11	.57	.51	.27	.91	.97
.72	.89	.12	.59	.52	.29	.92	22.99
.73	.90	.13	.61	.53	.30	.93	23.00
.74	.92	.14	.62	.54	.32	.94	.02
.75	.94	.15	.64	.55	.34	.95	.04
.76	.96	.16	.66	.56	.36	.96	.06
.77	.97	.17	.68	.57	.37	.97	.07
.78	20.99	.18	.69	.58	.39	.98	.09
.79	21.01	.19	.71	.59	.41	.99	.11

LABORATORY PROCEDURES FOR SEAWATER AND SHELLFISH

TABLE 3—(CONTINUED)

Cl/liter$_{(20)}$	S‰	Cl/liter$_{(20)}$	S‰	Cl/liter$_{(20)}$	S‰	Cl/liter$_{(20)}$	S‰
13.00	23.13	13.40	23.83	13.80	24.52	14.20	25.22
.01	.14	.41	.84	.81	.54	.21	.24
.02	.16	.42	.86	.82	.56	.22	.26
.03	.18	.43	.88	.83	.58	.23	.27
.04	.20	.44	.89	.84	.59	.24	.29
.05	.21	.45	.91	.85	.61	.25	.31
.06	.23	.46	.93	.86	.63	.26	.32
.07	.25	.47	.95	.87	.65	.27	.34
.08	.27	.48	.96	.88	.66	.28	.36
.09	.28	.49	.98	.89	.68	.29	.38
13.10	23.30	13.50	24.00	13.90	24.70	14.30	25.39
.11	.32	.51	.02	.91	.72	.31	.41
.12	.34	.52	.03	.92	.73	.32	.43
.13	.35	.53	.05	.93	.75	.33	.45
.14	.37	.54	.07	.94	.77	.34	.46
.15	.39	.55	.09	.95	.79	.35	.48
.16	.41	.56	.10	.96	.80	.36	.50
.17	.42	.57	.12	.97	.82	.37	.52
.18	.44	.58	.14	.98	.84	.38	.53
.19	.46	.59	.16	.99	.85	.39	.55
13.20	23.48	13.60	24.17	14.00	24.87	14.40	25.57
.21	.49	.61	.19	.01	.89	.41	.59
.22	.51	.62	.21	.02	.91	.42	.60
.23	.53	.63	.23	.03	.92	.43	.62
.24	.55	.64	.24	.04	.94	.44	.64
.25	.56	.65	.26	.05	.96	.45	.66
.26	.58	.66	.28	.06	.98	.46	.67
.27	.60	.67	.30	.07	24.99	.47	.69
.28	.62	.68	.31	.08	25.01	.48	.71
.29	.63	.69	.33	.09	.03	.49	.72
13.30	23.65	13.70	24.35	14.10	25.05	14.50	25.74
.31	.67	.71	.37	.11	.06	.51	.76
.32	.69	.72	.38	.12	.08	.52	.78
.33	.70	.73	.40	.13	.10	.53	.79
.34	.72	.74	.42	.14	.12	.54	.81
.35	.74	.75	.44	.15	.13	.55	.83
.36	.76	.76	.45	.16	.15	.56	.85
.37	.77	.77	.47	.17	.17	.57	.86
.38	.79	.78	.49	.18	.19	.58	.88
.39	.81	.79	.51	.19	.20	.59	.90

TABLE 3—(CONTINUED)

$Cl/\text{liter}_{(20)}$	$S‰$	$Cl/\text{liter}_{(20)}$	$S‰$	$Cl/\text{liter}_{(20)}$	$S‰$	$Cl/\text{liter}_{(20)}$	$S‰$
14.60	25.92	15.00	26.61	15.40	27.31	15.80	28.00
.61	.93	.01	.63	.41	.32	.81	.02
.62	.95	.02	.65	.42	.34	.82	.03
.63	.97	.03	.66	.43	.36	.83	.05
.64	25.99	.04	.68	.44	.38	.84	.07
.65	26.00	.05	.70	.45	.39	.85	.09
.66	.02	.06	.72	.46	.41	.86	.10
.67	.04	.07	.73	.47	.43	.87	.12
.68	.06	.08	.75	.48	.44	.88	.14
.69	.07	.09	.77	.49	.46	.89	.16
14.70	26.09	15.10	26.79	15.50	27.48	15.90	28.17
.71	.11	.11	.80	.51	.50	.91	.19
.72	.13	.12	.82	.52	.51	.92	.21
.73	.14	.13	.84	.53	.53	.93	.23
.74	.16	.14	.86	.54	.55	.94	.24
.75	.18	.15	.87	.55	.57	.95	.26
.76	.19	.16	.89	.56	.58	.96	.28
.77	.21	.17	.91	.57	.60	.97	.29
.78	.23	.18	.92	.58	.62	.98	.31
.79	.25	.19	.94	.59	.64	.99	.33
14.80	26.26	15.20	26.96	15.60	27.65	16.00	28.35
.81	.28	.21	.98	.61	.67	.01	.36
.82	.30	.22	26.99	.62	.69	.02	.38
.83	.32	.23	27.01	.63	.71	.03	.40
.84	.33	.24	.03	.64	.72	.04	.42
.85	.35	.25	.05	.65	.74	.05	.43
.86	.37	.26	.06	.66	.76	.06	.45
.87	.39	.27	.08	.67	.77	.07	.47
.88	.40	.28	.10	.68	.79	.08	.49
.89	.42	.29	.12	.69	.81	.09	.50
14.90	26.44	15.30	27.13	15.70	27.83	16.10	28.52
.91	.46	.31	.15	.71	.84	.11	.54
.92	.47	.32	.17	.72	.86	.12	.55
.93	.49	.33	.18	.73	.88	.13	.57
.94	.51	.34	.20	.74	.90	.14	.59
.95	.53	.35	.22	.75	.91	.15	.61
.96	.54	.36	.24	.76	.93	.16	.62
.97	.56	.37	.25	.77	.95	.17	.64
.98	.58	.38	.27	.78	.97	.18	.66
.99	.59	.39	.29	.79	.98	.19	.68

Table 3—(Continued)

Cl/liter$_{(20)}$	S‰	Cl/liter$_{(20)}$	S‰	Cl/liter$_{(20)}$	S‰	Cl/liter$_{(20)}$	S‰
16.20	28.69	16.60	29.39	17.00	30.08	17.40	30.77
.21	.71	.61	.40	.01	.09	.41	.79
.22	.73	.62	.42	.02	.11	.42	.80
.23	.75	.63	.44	.03	.13	.43	.82
.24	.76	.64	.45	.04	.15	.44	.84
.25	.78	.65	.47	.05	.16	.45	.85
.26	.80	.66	.49	.06	.18	.46	.87
.27	.82	.67	.51	.07	.20	.47	.89
.28	.83	.68	.52	.08	.22	.48	.91
.29	.85	.69	.54	.09	.23	.49	.92
16.30	28.87	16.70	29.56	17.10	30.25	17.50	30.94
.31	.88	.71	.58	.11	.27	.51	.96
.32	.90	.72	.59	.12	.28	.52	.98
.33	.92	.73	.61	.13	.30	.53	30.99
.34	.94	.74	.63	.14	.32	.54	31.01
.35	.95	.75	.65	.15	.34	.55	.03
.36	.97	.76	.66	.16	.35	.56	.04
.37	28.99	.77	.68	.17	.37	.57	.06
.38	29.00	.78	.70	.18	.39	.58	.08
.39	.02	.79	.71	.19	.41	.59	.10
16.40	29.04	16.80	29.73	17.20	30.42	17.60	31.11
.41	.06	.81	.75	.21	.44	.61	.13
.42	.07	.82	.77	.22	.46	.62	.15
.43	.09	.83	.78	.23	.47	.63	.17
.44	.11	.84	.80	.24	.49	.64	.18
.45	.13	.85	.82	.25	.51	.65	.20
.46	.14	.86	.84	.26	.53	.66	.22
.47	.16	.87	.85	.27	.54	.67	.23
.48	.18	.88	.87	.28	.56	.68	.25
.49	.20	.89	.89	.29	.58	.69	.27
16.50	29.21	16.90	29.90	17.30	30.60	17.70	31.29
.51	.23	.91	.92	.31	.61	.71	.30
.52	.25	.92	.94	.32	.63	.72	.32
.53	.26	.93	.96	.33	.65	.73	.34
.54	.28	.94	.97	.34	.66	.74	.36
.55	.30	.95	29.99	.35	.68	.75	.37
.56	.32	.96	30.01	.36	.70	.76	.39
.57	.33	.97	.03	.37	.72	.77	.41
.58	.35	.98	.04	.38	.73	.78	.42
.59	.37	.99	.06	.39	.75	.79	.44

TABLE 3—(CONTINUED)

Cl/liter$_{(20)}$	$S‰$	Cl/liter$_{(20)}$	$S‰$	Cl/liter$_{(20)}$	$S‰$	Cl/liter$_{(20)}$	$S‰$
17.80	31.46	18.20	32.15	18.60	32.84	19.00	33.53
.81	.48	.21	.17	.61	.86	.01	.54
.82	.49	.22	.18	.62	.87	.02	.56
.83	.51	.23	.20	.63	.89	.03	.58
.84	.53	.24	.22	.64	.91	.04	.60
.85	.55	.25	.23	.65	.92	.05	.61
.86	.56	.26	.25	.66	.94	.06	.63
.87	.58	.27	.27	.67	.96	.07	.65
.88	.60	.28	.29	.68	.98	.08	.67
.89	.61	.29	.30	.69	32.99	.09	.68
17.90	31.63	18.30	32.32	18.70	33.01	19.10	33.70
.91	.65	.31	.34	.71	.03	.11	.72
.92	.67	.32	.36	.72	.05	.12	.73
.93	.68	.33	.37	.73	.06	.13	.75
.94	.70	.34	.39	.74	.08	.14	.77
.95	.72	.35	.41	.75	.10	.15	.79
.96	.74	.36	.42	.76	.11	.16	.80
.97	.75	.37	.44	.77	.13	.17	.82
.98	.77	.38	.46	.78	.15	.18	.84
.99	.79	.39	.48	.79	.17	.19	.85
18.00	31.80	18.40	32.49	18.80	33.18	19.20	33.87
.01	.82	.41	.51	.81	.20	.21	.89
.02	.84	.42	.53	.82	.22	.22	.91
.03	.86	.43	.55	.83	.23	.23	.92
.04	.87	.44	.56	.84	.25	.24	.94
.05	.89	.45	.58	.85	.27	.25	.96
.06	.91	.46	.60	.86	.29	.26	.97
.07	.92	.47	.61	.87	.30	.27	33.99
.08	.94	.48	.63	.88	.32	.28	34.01
.09	.96	.49	.65	.89	.34	.29	.03
18.10	31.98	18.50	32.67	18.90	33.36	19.30	34.04
.11	31.99	.51	.68	.91	.37	.31	.06
.12	32.01	.52	.70	.92	.39	.32	.08
.13	.03	.53	.72	.93	.41	.33	.09
.14	.05	.54	.73	.94	.42	.34	.11
.15	.06	.55	.75	.95	.44	.35	.13
.16	.08	.56	.77	.96	.46	.36	.15
.17	.10	.57	.79	.97	.48	.37	.16
.18	.11	.58	.80	.98	.49	.38	.18
.19	.13	.59	.82	.99	.51	.39	.20

TABLE 3—(CONTINUED)

Cl/liter$_{(20)}$	S‰	Cl/liter$_{(20)}$	S‰	Cl/liter$_{(20)}$	S‰	Cl/liter$_{(20)}$	S‰
19.40	34.22	19.80	34.90	20.20	35.59	20.60	36.28
.41	.23	.81	.92	.21	.61	.61	.30
.42	.25	.82	.94	.22	.63	.62	.31
.43	.27	.83	.95	.23	.64	.63	.33
.44	.28	.84	.97	.24	.66	.64	.35
.45	.30	.85	34.99	.25	.68	.65	.36
.46	.32	.86	35.01	.26	.70	.66	.38
.47	.34	.87	.02	.27	.71	.67	.40
.48	.35	.88	.04	.28	.73	.68	.41
.49	.37	.89	.06	.29	.74	.69	.43
19.50	34.39	19.90	35.07	20.30	35.76	20.70	36.45
.51	.40	.91	.09	.31	.78	.71	.47
.52	.42	.92	.11	.32	.80	.72	.48
.53	.44	.93	.13	.33	.82	.73	.50
.54	.46	.94	.14	.34	.83	.74	.52
.55	.47	.95	.16	.35	.85	.75	.53
.56	.49	.96	.18	.36	.87	.76	.55
.57	.51	.97	.19	.37	.88	.77	.57
.58	.52	.98	.21	.38	.90	.78	.59
.59	.54	.99	.23	.39	.92	.79	.60
19.60	34.56	20.00	35.25	20.40	35.93	20.80	36.62
.61	.58	.01	.27	.41	.95	.81	.64
.62	.59	.02	.28	.42	.97	.82	.65
.63	.61	.03	.30	.43	35.99	.83	.67
.64	.63	.04	.32	.44	36.00	.84	.69
.65	.64	.05	.34	.45	.02	.85	.71
.66	.66	.06	.35	.46	.04	.86	.72
.67	.68	.07	.37	.47	.06	.87	.74
.68	.70	.08	.39	.48	.07	.88	.76
.69	.71	.09	.40	.49	.09	.89	.77
19.70	34.73	20.10	35.42	20.50	36.11	20.90	36.79
.71	.75	.11	.44	.51	.12	.91	.81
.72	.77	.12	.46	.52	.14	.92	.83
.73	.78	.13	.47	.53	.16	.93	.84
.74	.80	.14	.50	.54	.18	.94	.86
.75	.82	.15	.51	.55	.19	.95	.88
.76	.83	.16	.52	.56	.21	.96	.89
.77	.85	.17	.54	.57	.23	.97	.91
.78	.87	.18	.56	.58	.24	.98	.93
.79	.89	.19	.58	.59	.26	.99	.94

APPENDIX

TABLE 3—(CONCLUDED)

Cl/liter$_{(20)}$	$S‰$	Cl/liter$_{(20)}$	$S‰$	Cl/liter$_{(20)}$	$S‰$	Cl/liter$_{(20)}$	$S‰$
21.00	36.96	.25	.39	21.50	37.82	.75	.24
.01	36.98	.26	.40	.51	.83	.76	.26
.02	37.00	.27	.42	.52	.85	.77	.28
.03	.01	.28	.44	.53	.87	.78	.29
.04	.03	.29	.46	.54	.89	.79	.31
.05	.05	21.30	37.47	.55	.90	21.80	38.33
.06	.06	.31	.49	.56	.92	.81	.34
.07	.08	.32	.51	.57	.94	.82	.36
.08	.10	.33	.53	.58	.95	.83	.38
.09	.12	.34	.54	.59	.97	.84	.40
21.10	37.13	.35	.56	21.60	37.99	.85	.41
.11	.15	.36	.58	.61	38.00	.86	.43
.12	.17	.37	.59	.62	.02	.87	.45
.13	.18	.38	.61	.63	.04	.88	.46
.14	.20	.39	.63	.64	.06	.89	.48
.15	.22	21.40	37.65	.65	.07	21.90	38.50
.16	.24	.41	.66	.66	.09	.91	.51
.17	.25	.42	.68	.67	.11	.92	.53
.18	.27	.43	.70	.68	.12	.93	.55
.19	.29	.44	.71	.69	.14	.94	.57
21.20	37.30	.45	.73	21.70	38.16	.95	.58
.21	.32	.46	.75	.71	.17	.96	.60
.22	.34	.47	.77	.72	.19	.97	.62
.23	.36	.48	.78	.73	.21	.98	.63
.24	.37	.49	.80	.74	.23	.99	.65
						22.00	38.67

TABLE 4. CONVERSION OF CHLOROSITY C_o at 20°C TO CHOLORINITY Cl‰

Calculated Chlorosity	Subtract for Chlorinity	Calculated Chlorosity	Subtract for Chlorinity
9.95—10.35	−0.12	15.88—16.17	−0.31
10.36—10.75	−0.13	16.18—16.32	−0.32
10.76—11.15	−0.14	16.33—16.62	−0.33
11.16—11.46	−0.15	16.63—16.82	−0.34
11.47—11.76	−0.16	16.83—17.11	−0.35
11.77—12.06	−0.17	17.12—17.32	−0.36
12.07—12.46	−0.18	17.33—17.57	−0.37
12.47—12.86	−0.19	17.58—17.82	−0.38
12.87—13.07	−0.20	17.83—18.02	−0.39
13.08—13.37	−0.21	18.03—18.27	−0.40
13.38—13.67	−0.22	18.28—18.47	−0.41
13.68—14.02	−0.23	18.48—18.67	−0.42
14.03—14.27	−0.24	18.68—18.97	−0.43
14.28—14.52	−0.25	18.98—19.17	−0.44
14.53—14.82	−0.26	19.18—19.32	−0.45
14.83—15.09	−0.27	19.33—19.52	−0.46
15.10—15.37	−0.28	19.53—19.77	−0.47
15.38—15.68	−0.29	19.78—19.97	−0.48
15.69—15.87	−0.30		

QR 106 .A54 1984
American Public Health
 Association.
Laboratory procedures for
 the examination of seawater

APR 1 0 1989